T0192356

Software-Defined Networking and Security

From Theory to Practice

Data-Enabled Engineering Series

Series Editors: Nong Ye, Arizona State University, Phoenix, USA

Convolutional Neural Networks in Visual Computing

A Concise Guide
Ragav Venkatesan and Baoxin Li

For more information about this series, please visit: www.crcpress.com/Data-Enabled-Engineering/book-series/CRCDATENAENG

Software-Defined Networking and Security

From Theory to Practice

Dijiang Huang
Ankur Chowdhary
Sandeep Pisharody

CRC Press
Taylor & Francis Group
Boca Raton London New York

CRC Press is an imprint of the
Taylor & Francis Group, an **informa** business

CRC Press
Taylor & Francis Group
6000 Broken Sound Parkway NW, Suite 300
Boca Raton, FL 33487-2742

First issued in paperback 2020

ISBN-13: 978-0-8153-8114-3 (hbk)
ISBN-13: 978-0-367-78064-7 (pbk)

Library of Congress Cataloging-in-Publication Data

Names: Huang, Dijiang, author. | Chowdhary, Ankur, author. | Pisharody, Sandeep, author.
Title: Software-defined networking and security : from theory to practice / Dijiang Huang, Ankur Chowdhary and Sandeep Pisharody.
Description: First edition. | Boca Raton, FL : CRC Press/Taylor & Francis Group, 2018. | Series: Data-enabled engineering
Identifiers: LCCN 2018037448 | ISBN 9780815381143 (hardback : acid-free paper)
Subjects: LCSH: Software-defined networking (Computer network technology)
Classification: LCC TK5105.5833 .H83 2018 | DDC 005.8--dc23
LC record available at https://lccn.loc.gov/2018037448

Visit the Taylor & Francis Web site at
http://www.taylorandfrancis.com

and the CRC Press Web site at
http://www.crcpress.com

To Lu, Alex, and Sarah: love,

Dijiang/Dad

To my mother, father, and sister, Vaishali

Ankur

To Shuchi, you mean the world to me!

Sandeep

Contents

Preface

Why This Book?

The project of working on a book including both education and research content on security of Software-Defined Networking (SDN) and Network Functions Virtualization (NFV) has been laid out since 2015, when the first author was awarded the National Science Foundation (NSF) research award on the Secure and Resilient Networking (SRN) project [125]. Before the SRN award, significant effort was also put together by the Secure Networking and Computing (SNAC) research group at Arizona State University (ASU) in the area of computer network security. The book's contents are mainly built on the research and education efforts of book authors and their colleagues in SNAC. Encouraged by Professor Ye Nong, who is a book series editor at CRC Press, the authors decided to work on the book in early 2017.

Security in the SDN/NFV domain is a fast-evolving area. It is impossible for us to provide a thorough description to cover all aspects of security in SDN/ NFV. Generally speaking security in SDN/NFV should include both SDN/ NFV security and how to use SDN/NFV to provide stronger security features for computer networks. This book focuses on the latter, particularly on how to use the programmability capability of SDN/NFV to establish a Moving Target Defense (MTD) network security infrastructure.

SDN is an emerging research area that has attracted a lot of attention from academia, industry, and government. SDN packs in itself immense possibilities from supporting the creation of consolidated data centers and better load balancing, to seamless mobility and secure networks. It is an innovation that allows us to control and program the network in a way to make it responsive to networking events, for example, events caused by security breaches, in a more proactive fashion. This book seeks to enlighten and educate the reader about cyber maneuvers or adaptive and intelligent cyber defense. Prior to implementing these proactive cyber defense techniques, it is important to analyze potential threats in an environment, detect attacks, and implement countermeasures in a way that expends attacker resources while preserving user experience. This book discusses theory and tools to detect multi-stage attacks and how to use SDN/NFV approaches to build a more proactive approach and evaluate their effectiveness.

The SDN approach separates control and data planes which improves optimization of network policies, and by providing easy access to flow tables, it gives a real-time control over the network switches, allowing administrators to monitor and control the route of packets flowing through the network. Thus, the packets, which otherwise flow according to fixed and firmware-defined security rules, can now be analyzed and controlled according to dynamic user-defined rules. This traffic reshaping capability of SDN promises further developments in networking and allows exploitation of a true control over the traffic. One of the many applications of SDN can be in improving the security by controlling traffic flow in the network by redirecting the packets from a suspicious node to an inspection node where in-depth examination of these packets can be performed.

SDN can also help in implementation of other techniques for improving security in a Software-Defined Infrastructure (SDI) cloud environment, such as reconfiguring the network dynamically to enforce packet forwarding, blocking, redirection, reflection, changing of MAC or IP address, limiting the packet flow rate, etc. These solutions can be considered less intrusive alternatives to security countermeasures taken at the host level, and offer centralized control of the distributed network. Continuing with this notion of security with SDN, in this book, we will introduce the basic features of SDN technologies and explain how to deploy a secure cloud computing system based on SDN solutions.

The development of SDN/NFV and their supported security solutions have been shifted from implementing static SDN/NFV-supported security functions or appliance to a more intelligent SDN/NFV security control framework to intelligently and efficiently managing and deploying SDN/NFV security features considering various factors such as effectiveness on countering attacks, intrusiveness to good users, and system cost to deploying ad hoc security functions. With the research and development in the area of SDN security in the past few years, we strongly feel that SDN security has evolved beyond the networking area. A more general area, usually called software-defined and programmable systems, has emerged including every aspect of a computing and networked system.

Audience

Our goal has been to create a book that can be used by a diverse set of audiences, and with varied levels of background. It can serve as a reference book for college students, instructors, researchers, and software developers interested in developing SDN/NFV-supported applications. It can serve as a good reference book for undergraduate and graduate courses focusing on computer network and security. Specifically, we set out to create a book that can be used by professionals as well as students and researchers. In general, this is intended as a self-study. We assume that the reader already has some basic knowledge of computers, networking, and software-based service models. Among professionals, the intent has been to cover two broad groups: computer networking and security developers including SDN/NFV and system security, with the overall goal to bring out issues that one group might want to understand that the other group faces.

For students, this book is intended to help learn about virtualization for SDN/NFV in depth, along with lessons from security aspects, and operational and implementation experience. For researchers who want to know what has been done so far and what are the critical issues to address for the next generation of computer network security, this is intended as a helpful reference.

Organization and Approach

The organization of the book is summarized as follows.

In the first part of this book, we will focus on the foundation of computer networks. Particularly, we first present basic concepts of computer networks in Chapter 1, covering layered network architecture, services, and packet encapsulation. Addresses such as MAC addresses, IP addresses, and port numbers play very important roles in network layers, and thus we put a lot of emphasis on these fundamental concepts for novel readers in computer network areas. In order to help understand computer networks, we provide a comprehensive description for relations between physical networks, logical networks, and virtual networks. Moreover, several important computer network services such as ARP, DHCP, DNS, and NAT are presented. Finally, network routing in IP networks and software-defined networks are described. In summary, this chapter provides a foundation for novice readers in the networking area, and the presented concepts and descriptions are necessary to understand the rest of this book.

Both SDN and NFV are heavily built on network virtualization technologies. Thus, in Chapter 2, we provide a comprehensive view of existing network virtualization solutions. This chapter starts with basic virtualization concepts and technologies including OpenFlow, OPNFV, and virtual networking embedding problems. Then, layer-2 virtual networking solutions such as Linux bridge and open virtual switches are described in details. Tunneling solutions such as VLAN, VxLAN, and GRE are widely used to build existing virtual and programmable networks, which are also presented. Finally, virtual routing and forwarding are described. Understanding materials provided in this chapter is critical to understanding modern computer networking solutions.

In Chapter 3, SDN and NFV are introduced and described in more detail. Motivation for both these paradigms, their benefits, challenges, and use cases are described. Leading frameworks that implement NFV and SDN are also discussed. This is followed by a discussion of the symbiotic nature of these two paradigms and their interworking. Finally, an introduction of P4 and PISA is presented as an advanced topic for deep programmable SDN.

This book focuses on security aspects of using SDN and NFV solutions. To understand the basics of computer network security, in Chapter 4, we first describe several important security concepts to understand differences between threat model and attack model, defense in depth, cyber killer chain, and their limitations. Then, several network exploration approaches such as network mapping, port scanning, vulnerability scanning, and penetration testing are presented. Later, we focus the discussion on preventive techniques such as firewalls and intrusion prevention. Next, detection and monitoring techniques are presented. Finally, we briefly discuss what is network security assessment. All the presented basic security concepts and mechanisms build the fundamental network security services and they can be implemented in an SDN/NFV networking environment and controlled by using the programmable features.

Finally, in the first part of Chapter 5 we analyze the threat model and attack vectors that are part of traditional networks and new threats that are introduced as a part of an SDN/NFV framework. As part of NFV security we discuss intra-VNF, extra-VNF security threats, and countermeasures that can help in addressing NFV security challenges. The SDN security has been distributed into threat vectors targeting different layers of SDN infrastructure, i.e., SDN data plane and control plane. Additionally, we discuss challenges specific to SDN, OpenFlow protocol (one of the most common SDN protocols), OpenFlow switch, and attack countermeasures to deal with SDN threats.

In the second part of this book, we will focus on the advanced topics on how to build a secure networking solution by utilizing SDN and NFV, particularly, the presentation focus on a new MTD concept, which utilizes the programmability capability of SDN to automate the defense-in-depth control in terms of security monitoring and analysis, security countermeasure services selection, and deployment. Advanced topics such as security policy management and machine learning models involving SDN/NFV are also presented in this part.

In Chapter 6, we present the security microsegmentation concept and its realization, which is illustrated by using VMware's NSX service model. We explain how a security service model such as firewall is transited to a more scrutinized approach, i.e., microsegmentation. A highly related security service – distributed firewall – is discussed in detail followed by illustration of microsegmentation concepts. Finally, a microsegmentaion case study is discussed in detail. Microsegmentation is a good starting point to improve the programmability and agility of network security services. It is the trend; however, it faces high network and security state management challenges to be overcome to make microsegmentation practical.

In Chapter 7, we discuss proactive security mechanisms to reduce the attack surface and limit the capabilities of the attacker, as compared to a static defense mechanism, where the attacker has asymmetric advantage of time. We discuss different MTD techniques, such as random host mutation, port hopping, and their impact on the service availability and resources (compute and storage) in the network. SDN helps in deployment of different MTD techniques in an automated fashion, hence we have dedicated one section of this chapter to SDN-based MTD. We analyze the attack-defense scenarios as a dynamic game between the attacker and the defender, and evaluate different MTD frameworks from a qualitative and quantitative perspective in this chapter.

Chapter 8 is dedicated to cybersecurity metrics utilized for quantification of the attack. The most common metric Common Vulnerability Scoring System (CVSS) has been discussed in detail in the first section of this chapter. We dedicate the rest of the chapter to attack graphs and attack trees, which are used to represent the dependencies between network services and vulnerabilities. The attack representation methods discussed in this chapter help in representing multi-hop attacks and possible countermeasures (attack countermeasure trees) in a simplified and intuitive fashion. We also address the limitations associated with different ARMs in this chapter.

The end-to-end delivery of network traffic requires the packets to be processed by different security and optimization Virtual Network Functions (VNFs), such as application firewall, load balancer, etc. This chaining of VNFs—Service Function Chaining (SFC)—is discussed in Chapter 9. The key challenges associated with incorporation of SFC and the role of SDN as an enabler of SFC, in identification of dependencies between VNFs and compilation into policy aware SFC, have been discussed in this chapter. The research and industry SDN/NFV-based SFC testbeds and their architecture have been discussed in detail. Additionally, policy-aware SFC and secured SFC have been discussed with illustrative examples in this chapter.

In Chapter 10, policy conflicts in security implementations are discussed, with emphasis on flow rule conflicts in SDN environments. A formalism for flow rule conflicts in SDN environments is described. A conflict detection and resolution model is discussed that ensure no two flow rules in a distributed SDN-based cloud environment have conflicts at any layer; thereby assuring consistent conflict-free security policy implementation and prevention of information leakage.

Chapter 11 is dedicated to analysis of advancements in the fields of intelligent security, such as application of Machine Learning (ML) and Artificial Intelligence (AI) in the field of cybersecurity, which use cases such as the role AI can play in improvement of a current Intrusion Detection System (IDS). We discuss the SDN-based intelligent network security solution that can incorporate these advancements in the field of intelligent cybersecurity. We discuss the Advanced Persistent Threats (APTs) and different stages of APTs in the cyberkill chain, along with suggestions for mitigation of APT using SDN-enabled microsegmentation and secured SFC. In the last section of this chapter, we discuss some key challenges that limit the application of ML and AI in cybersecurity, such as high cost of errors, semantic gap, highly variant network traffic, etc.

Bonus Materials and Online Resources

The book, in its printed form, has 11 chapters; additional support materials, for example, source codes, implementation instructions, additional reading, and instructional materials.

Dijiang Huang
Tempe, Arizona, USA
Dijiang.Huang@asu.edu

Ankur Chowdhary
Tempe, Arizona, USA
achaud16@asu.edu

Sandeep Pisharody
Lexington, Massachusetts, USA
Sandeep.Pisharody@ll.mit.edu

Acknowledgments

This book cannot be done without the tireless and hard work of three authors who have collaboratively contributed the materials and presentation for each chapter. However, each chapter also has a lead author to coordinate and organize the materials and presentations. Particularly, Dr. Dijiang Huang leads Chapters 1, 2, 4, and 6; Ankur Chowdhary leads Chapters 5, 7, 8, 9, and 11; and Dr. Sandeep Pisharody leads Chapters 3 and 10. Dr. Chun-Jen Chung and Yuli Deng help the authors establish the SDN/NFV testing platform and source code testing. Some parts of MTD and APT solutions are derived from Adel Alshamrani and Sowmya Myneni's research work.

Dr. Dijiang Huang leads the Secure Networking and Computing (SNAC) group at Arizona State University (ASU). SNAC is formed by graduate students who are working on various research and development projects in the areas of computer networking security, cloud computing security, applied cryptography, and IoT security, etc. SNAC hosts regular meetings for group members to share research results and discuss research issues. Most SDN security-related work has been studied through SNAC meetings. The authors also gratefully thank current SNAC members: Adel Alshamrani, Chun-Jen Chung, Yuli Deng, Qiuxiang Dong, Jiayue Li, Fanjie Lin, Duo Lu, Sowmya Myneni, Abdulhakim Sabur, and Zeng Zhen, who greatly inspired them through discussions, seminars, and project collaborations.

The authors would also like to thank SNAC alumni: Abdullah Alshalan (King Saud University), Bing Li (Google), Zhijie Wang (General Electric Research Laboratory), Tianyi Xing (Wal-Mart Research Lab), Zhibin Zhou (Huawei), Yang Qin (Facebook), Bhakti Bohara (Akamai Technologies), Shilpa Nagendra (Commvault Systems), Janakarajan Natarajan (AMD), Qingyun Li, Pankaj Kumar Khatkar (CAaNES), Zhenyang Xiong (Omedix), Ashwin Narayan Prabhu Verleker (Microsoft), Xinyi Dong (Amazon), Yunji Zhong (Microsoft), Aniruddha Kadne (F5), Sushma Myneni (Microchip), Nirav Shah (Intel), Vetri Arasan (Garmin), Le Xu (Microsoft), Oussama Mjihil (Fulbright scholar student, Hassan I University, Morocco), Iman Elmir (Hassan I University, Morocco), Xiang Gu (Nantong University, China), Bo Li (Yunnan University, China), Zhiyuan Ma (UESTC, China), Weiping Peng (Henan Polytechnic University, China), Jin Wang (Nantong University, China), Aiguo Chen (University of Electronic Science and Technology, China), Jingsong Cui (Wuhan University, China), Weijia Wang (Beijing Jiaotong University, China), and Chunming Wu (Southwest University, China).

Special thanks to the NSF SRN project collaborators Professor Deep Medhi, Professor Kishor Trivedi, and their research group members based on our research collaborations, and useful discussions through group meetings and various collaborative private communication.

Our immediate family members suffered the most during our long hours of being glued to our laptops. Throughout the entire duration, they provided all sorts of support, entertainment, and "distractions." Dijiang would like to thank his wife, Lu, and their son, Alexander, and daughter, Sarah, for love and patience, and for enduring this route. He would also like to thank his father, Biyao, mother, Zhenfen, and brother, Dihang, for their understanding when he was not able to call them regularly and postponing several trips to China for a family reunion. He would also like to acknowledge his family members, Shan and Yicheng. Finally, he would like to thank his many friends for numerous support.

Ankur would like to thank his father, Vikram Singh, mother, Munesh, sister, Vaishali, all his cousins, and friends, Tarun, Abhishek, Puneet, Rahul, Chandan, Sunit, Prashant, Rushang, for their support and guidance.

Sandeep would like to express his profound gratitude to his family for being a pillar of support through this work. His amazing wife, Shuchi, shouldered far more than her fair share of parenting two kids and household responsibilities, while he sat zoned out in front of a computer. His parents and brother were the perfect role models throughout childhood and now adulthood. And finally, his children, Gayathri and Om, were his compass and helped put things into perspective.

About the Authors

LinkedIn QR Code

Dr. Dijiang Huang graduated in telecommunications from Beijing University of Posts and Telecommunications (China) with a bachelor's degree in 1995; his first job was a network engineer in the computer center of Civil Aviation Administration of China (CAAC), where he received four years of industry working experience. He then came to the University of Missouri-Kansas City (UMKC) in the US to pursue his graduate study in the joint computer networking and telecommunication networking program of computer science; and he earned an MS and PhD in computer science in 2001 and 2004, respectively. During his study at UMKC, he became interested in the research areas of mobile computing and security, and focused his research on network security and mobile networks.

After graduating with a PhD, Dr. Huang joined the Computer Science and Engineering (CSE) department at Arizona State University (ASU) as an assistant professor. One of his early research areas was securing MANET communication and networking protocols. Later, he realized that the cross-layer approach was extremely important in making a MANET solution more efficient and practical. Gradually, he looked into the research problem of how to build a situation-aware solution that better supported MANET applications, considering various unstable issues due to node mobility and intermittent communication. Considering mobiles trying to utilize all reachable resources to support their applications, this situation is very similar to the resource management scenario for cloud computing, of course with a different context, running environment, programming, and virtualization capabilities and constraints.

In 2010, Dr. Huang was awarded the Office of Naval Research (ONR) Young Investigator Program (YIP) award for a research project to establish a secure mobile cloud computing system. The main task of the award was to develop a secure and robust mobile cloud and networking system to support trustworthy mission-critical operations and resource management, considering communication, networking, storage, computation, and security requirements and constraints. The boom of Software Defined Networking (SDN) has changed the playground of computer network security, which has become more dynamic, automatic, and intelligent in the past few years. His research has been focused on a more intelligent and MTD by incorporating dynamic learning models into security analysis and decisions. This book can share his past research and development outcomes and provide a starting point to ride on the next research and development wave for software-defined security, which can benefit both research communities and practitioners.

Dr. Huang is currently a Fulton Entrepreneur Professor in the School of Computing Informatics Decision Systems Engineering (CIDSE) at ASU, he has published four US patents, and is a co-founder of two start-up companies: Athena Network Solutions LLC (ATHENETS) and CYNET LLC. He is currently leading the Secure Networking and Computing (SNAC) research group. Most of his current and previous research is supported by federal agencies such as National Science Foundation (NSF), ONR, Army Research Office (ARO), Naval Research Lab (NRL), National Science Foundation of China (NSFC), and North Atlantic Treaty Organization (NATO); and industries such as the Consortium of Embedded System (CES), Hewlett-Packard, and China Mobile. In addition to the ONR Young Investigator Award, he was also a recipient of the HP Innovation Research Program (IRP) Award, and the JSPS Fellowship. He is a senior member of IEEE and member of ACM. For more information about his research publications, teaching, and professional community services, please refer to http://www.public.asu.edu/~dhuang8/.

LinkedIn QR Code

Ankur Chowdhary is a PhD student at Arizona State University (ASU), Tempe, AZ, US. He received a B.Tech in information technology from GGSIPU, Delhi, India in 2011 and an MS in computer science from ASU in 2015. He has worked as an information security researcher for Blackberry Ltd. (2016), RSG (2015), and as an application developer for CSC Pvt 2011-2013 Ltd. His research interests include SDN, cloud security, web security, and application of machine learning in the field of security. He has co-authored 12 research publications, most of which are highly related to the subject matter of the book. He is highly involved in cybersecurity education and training for undergraduate and graduate students at ASU. He has been captaining the ASU Cybersecurity Defence Competition (CCDC) team from 2015 to the present. He is co-founder and CEO of CyNET LLC, a cybersecurity startup, whose prime objective is to provide SDN-based proactive security solutions for cloud networks and data centers. For more information about his research publications, teaching, and professional community services, please refer to http://www.public.asu.edu/~achaud16/.

ORCID QR Code

Dr. Sandeep Pisharody graduated from the University of Nebraska - Lincoln, with a BS in both electrical engineering and computer engineering in 2004. He went on to get an MS in electrical engineering in 2006 from the University of Nebraska - Lincoln, specializing in material sciences. He joined the industry as a network/security engineer in 2006, and spent the next eight years working in various capacities for Sprint Corp., Iveda Corp., Apollo Education Group, Insight, and the University of Phoenix. He went back to school under the guidance of the first author in 2013, and graduated with a PhD in computer science from Arizona State University. His areas of interest are network security, software-defined networks, cloud security, modeling and simulation, and human factors in security. He currently works as a technical staff member at MIT Lincoln Laboratory, Lexington, MA.

Part I

Foundations of Virtual Networking and Security

This book focuses on Network Functions Virtualization (NFV), Software Defined Networking (SDN), and security models built into/on SDN/NFV technologies. In this part, we will provide preliminary foundations of computer networking and NFV/SDN, which will help readers to understand advanced topics in this book. Moreover, the first part can be used as learning and teaching materials for students starting to learn computer and network security. Before moving forward, several important and highly related terms need to be clearly understood.

What is *Software?*

The Institute of Electrical and Electronics Engineers (IEEE) defines software as "The complete set of computer programs, procedures, and possibly associated documentation and data designated for delivery to a user" [225]. It possesses no mass, no volume, and no color, which makes it a non-degradable entity over a long period. Software does not wear out or get tired. In short, software can be simply defined as *a collection of programs, documentation and operating procedures.*

Software controls, integrates, and manages the hardware components of a computer system. It also instructs the computer what needs to be done to perform a specific task and how it is to be done. In general, software characteristics are classified into six major components:

1. *Functionality*, which refers to the degree of performance of the software against its intended purpose.
2. *Reliability*, which refers to the ability of the software to provide desired functionality under the given conditions.
3. *Usability*, which refers to the extent to which the software can be used with ease.
4. *Efficiency*, which refers to the ability of the software to use system resources in the most efficient manner, with the least waste of time and effort.
5. *Maintainability*, which refers to the ease with which modifications can be made in a software system to extend its functionality, improve its performance, or correct errors.
6. *Portability*, which refers to the ease with which software developers can transfer software from one platform to another, without (or with minimum) changes. In simple terms, it refers to the ability of software to function properly on different hardware and software platforms without making any changes in it.

In addition to the above characteristics, robustness and integrity are also important. Robustness refers to the degree to which the software can keep on functioning in spite of being provided with invalid data, while integrity refers to the degree to which unauthorized access to the software or data can be prevented.

Computers work only in response to instructions provided externally. For example, software instructs the hardware how to print a document, take input from the user, and display the output. Usually, the instructions to perform some intended tasks are organized into a program using a programming language like C, C++, Java, etc., and submitted to computer. The computer interprets and executes these instructions and provides response to the user accordingly. The set of programs intended to provide users with a set of interrelated functionalities is known as a software package.

What is *Software-Defined*?

The term *Software-Defined* is the ability to abstract the management and administrative capabilities of the technology. In terms of computer networking, it is the ability to control the provisioning of network devices, Virtual LANs (VLANs), Firewall rules, traffic engineering and Quality of Services

(QoS), etc. In summary, a system regarded as *Software-Defined* should rely on software to achieve the following interrelated and supported Abstraction, Automation and Adjustment *(AAA)* features:

Abstraction of Physical Resources: A software-defined system usually provides a set of Application Platform Interfaces (APIs) to abstract their physical resources to simplify its resource management and allocation.

Automation of Actions/Controls: Actions and controls can be executed based on the incorporated complex application logics by examining software running conditions or being triggered by software monitored systems with a certain level of adaptability and intelligence.

Adjustment of Configurations (Reconfiguration): Adjustments of system resource can be achieved through predictive configuration or control of workloads, which can be performed by changing past administrator defined rule sets.

Definitions of Software-Defined Systems

Software-Defined Networking (SDN) is an emerging architecture that is dynamic, manageable, cost-effective, and adaptable, making it ideal for the high-bandwidth, dynamic nature of today's applications. This architecture decouples the network control and forwarding functions enabling the network control to become directly programmable and the underlying infrastructure to be abstracted for applications and network services. The OpenFlow protocol is a foundational element for building SDN solutions.

Software-Defined Storage (SDS) is a computer program that manages data storage resources and functionality and has no dependencies on the underlying physical storage hardware. SDS is most often associated with software products designed to run on commodity server hardware. It enables users to upgrade the software separately from the hardware. Common characteristics of SDS products include the ability to aggregate storage resources, scale out the system across a server cluster, manage the shared storage pool and storage services through a single administrative interface, and set policies to control storage features and functionality.

Software-Defined Data Centers (SDDC) is often referred as to a data center where all infrastructure is virtualized and delivered as a service. Control of the data center is fully automated by software, meaning hardware configuration is maintained through intelligent software systems. This is in contrast to traditional data centers where the infrastructure is typically defined by hardware and devices. SDDCs are considered by many to be the evolution of virtualization and cloud computing as it provides a solution to support both legacy enterprise applications and new cloud computing services.

Software-Defined Power (SDP) is a solution to application-level reliability issues being caused by power problems, and it is about creating a layer of abstraction that makes it easier to continuously match resources with changing applications/services' needs, where the resource is the electricity required to power all of the equipment. Under the SDDC, the overall reliability of SDP is improved by shifting applications to a data center with the most dependable, available and cost-efficient power at any given time, which is implemented using a software system capable of combining IT and facility/building management systems, and automating standard operating procedures, resulting in the holistic allocation of power within and across data centers, as required by the ongoing changes in application load.

Software-Defined Infrastructure (SDI) is the definition of technical computing infrastructure entirely under the control of software with no operator or human intervention. It operates independent of any hardware specific dependencies and is programmatically extensible. In the SDI approach, an application's infrastructure requirements are defined declaratively (both functional and non-functional requirements) such that sufficient and appropriate hardware can be automatically derived and provisioned to deliver those requirements.

Software-Defined Everything (SDE) refers to various systems controlled by advanced software programs and constructed in a virtual, versus physical, hardware space. SDE is also often used as several technologies under one umbrella. For example, SDN involves the creation of virtualized networks, where physical hardware is replaced by a sophisticated software system. A SDDC uses virtualization techniques to construct a data center. SDS involves replacing a distributed hardware system with virtual storage systems. SDE also is a comprehensive idea based on new applications of technology, as SDE systems can be used to provide fully virtualized IT systems.

Software-Defined Virtual Networking (SDVN)

Network Functions Virtualization (NFV) is a concept of virtual networking architecture based on virtualization solutions to virtualize entire classes of network node functions into building blocks that may connect, or chain together, to create communication and networking services. A Virtual Network Function (VNF) may consist of one or more Virtual Machines (VMs) or containers running different software and processes, on top of standard high-volume servers, switches and storage devices, or even cloud computing infrastructure, instead of having custom hardware appliances for each network function.

NFV and SDN are commonly used together, and both move toward network virtualization and automation. However, they are different and serve different goals. An SDN can be considered a series of networking entities, such as switches, routers, and firewalls that are deployed in a highly automated manner. The automation may be achieved by using commercial or open source tools, like SDN controllers and OpenFlow based on the administrator's requirements. A full SDN may cover only relatively straightforward networking requirements, such as VLAN and interface provisioning.

NFV is the process of moving services like load balancing, firewalls and intrusion prevention systems away from dedicated hardware into a virtualized environment. Functions like caching and content control can easily be migrated to a virtualized environment, but they will not necessarily provide any significant reduction in operating costs until some intelligence is introduced. This is because a straight physical to virtual migration, from an operational perspective, achieves little beyond the initial reduction in power and rack-space consumption. Until some dynamic intelligence is introduced with an SDN technology, NFV network deployments inherit many of the same constraints as traditional hardware appliance deployments, such as static, administrator defined and managed policies.

In many intelligent networking use cases, SDN is linked to server virtualization, providing the glue that makes virtual networks stick together. This may involve NFV, but not necessarily. Thus, NFV and SDN are complementary technology initiatives. NFV moves services to a virtual environment but does not include policies to automate the environment. When SDN is combined with Network Functions Virtualization Infrastructure (NFVI), however, SDN's centralized management function can forward data packets from one network device to another, while NFV allows routing control functions to run on a VM or a container located in a rack mount server, for example. In essence, NFV and SDN make the network itself *programmable,* offering the promise of rapid innovation of network services customized and tightly integrated with specific application domains. The results of NFV/SDN research and development are creating fundamentally new measurement challenges in network behavior, software quality, and security properties of dynamically composed, programmable networks. Given the critical position of basic network control systems, the need to accurately measure and thoroughly test the safety, robustness, security and performance of SDN will be paramount in ensuring the success of these technologies' use in future missions or business-critical networks.

In this book, we refer to NFV and SDN technologies collectively as Software-Defined Virtual Networking (SDVN). They build the foundation to support computer network security functions in a virtual and highly autonomous networking environment to meet applications' needs and counter challenging malicious attacks.

Industry and academic leaders started the SDVN movement to change the economics and complexity of network innovation. Virtualized networking to support vast data centers was the initial commercial force driving SDVN, with network switch, Hypervisor, and cloud service vendors driving the pace and the direction of innovation. The realization of the power and potential of "opening up" networking platforms and enabling the seamless integration of programmable networks and applications set off a series of billion-dollar acquisitions and triggered even broader efforts by the industry to commoditize network hardware platforms and software environments. Today, the potential applications software-defined virtual networks range from global telecommunications to completely software-defined data centers. Current market analyses project the NFV/SDN market to reach $100B by 2020 [204].

1

Introduction of Computer Networks

When I, Dijiang, teach my computer network security class, I always pose a question: *"On the university campus, you boot up a laptop, open an email application, compose an email, provide the receiver's email address (e.g., john@xyz.com) and other related information such as email subject, make sure the email content and everything else is fine, and finally click send. During this procedure, what networking and application protocols have been invoked and what magic happened in the computer network to allow John to receive and view your email?"*.

To answer the question, I usually use a real-world example of *"sending a mail"* to emulate what may have happened in the virtual world, i.e., the Internet. Instead of sending an email, I can send a regular mail to John with the added limitation of a long mail delivery time. Before sending a regular mail, I need to do some preparation work such as putting the written letter into an envelope, writing down both sender's and receiver's addresses on the envelope, sealing it, and finally adding the requisite postage on the envelope. This letter preparation procedure is quite similar to invoking an email application to write an email before clicking the "send" button. In order to provide similar data privacy protection, i.e., using an envelope, I should use a secure email service, e.g., using an email client with encryption/decryption capability on both sides or using secure sockets layer (SSL) protocol to secure the communication links from my computer to John's.

Dropping the letter into a mailbox of a local post office is equivalent to clicking on the "send" button on the email client. The letter will be usually first sorted at the local post office based on the receiver's address to decide when and how to deliver (i.e., put the letter in a scheduled delivery truck to the next delivery hub), in which the sorting is based on the "scope" of the destination: international vs. domestic, city and street names or a zip code. Once the letter arrives at the destination's local post office, the letter distributor will check the receiver's home number and street name for the final delivery. This procedure is similar to what the Internet Protocol (IP)-based packet switching network uses to deliver an end-to-end packet that encapsulates the email content. Routers serve as counterparts for post offices and mail delivery hubs, while the transmission protocol serves the purpose of the mail delivery truck. In an IP packet, source IP address and destination IP address serve the purpose of sender's address and receiver's address for mail delivery, respectively. An IP address has a two-level data structure: the network address and host address, which can be mapped to the zip code, city, and street names; and home number, respectively. Internet routers only look at the network address to deliver the message to the destination's local network, and once arrived, the host address is used to deliver the message to the destination host.

The above example tells us that understanding how the Internet works is similar to understanding how our mail system works. Many of the existing Internet protocols help realize real-life applications or workflows. As a result, we can view the Internet is a virtual realization of our physical world. This can greatly help us understand how the Internet works and how we design a new computer networking solution.

In this chapter, we first present the foundations of computer networks focusing on packet switching networks in Section 1.1. Details will be provided to explain what addresses will be used at different layers of the TCP/IP protocol stack in Section 1.2. In Section 1.3, we will present basic concepts to understand physical and virtual networks; several important inter-networking protocols and services are illustrated in Section 1.4; and finally, IP network routing is introduced in Section 1.5.

1.1 Foundations of Computer Networks

In this section, we first introduce several important concepts in order to understand computer networks.

1.1.1 Protocol Layers

The Open Systems Interconnection (OSI) model is a conceptual model that characterizes and standardizes the communication functions of a tele-communication or computing system without regard to its underlying internal structure and technology. OSI model is standardized by the International Organization for Standardization (ISO) and International Electrotechnical Commission (IEC) with the identification ISO/IEC 7498-1. The goal of OSI is the interoperability of diverse communication systems with standard protocols. The model partitions a communication system into abstraction layers. Shown in Figure 1.1, on the right side is the original version of the model defined using seven layers, namely Physical, Data Link, Network, Transport, Session, Presentation, and Application. The corresponding TCP/IP protocol layers, which is the Internet Engineering Task Force (IETF) standard is presented on the left side of the Figure 1.1. The OSI model is conceptually sound to nicely present the relations among multiple protocols and their inter-dependencies. However, the simplicity of the TCP/IP protocol framework won the competition and has been widely adopted by many networking systems, especially, the Internet.

For both OSI and TCP/IP, a layer serves the layer above it and is served by the layer below it. For example, a layer that provides error-free

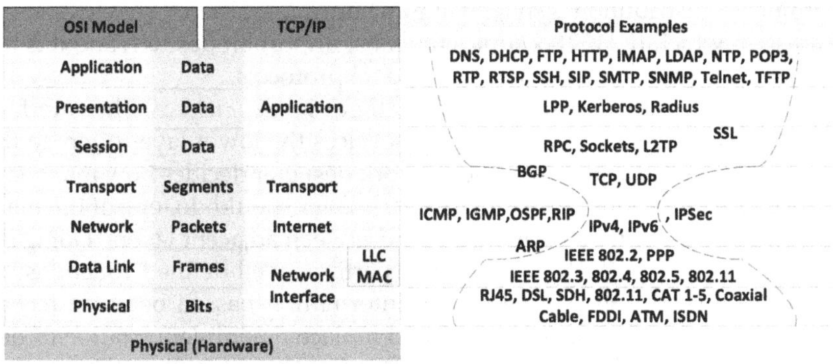

OSI Model		TCP/IP		Protocol Examples
Application	Data			DNS, DHCP, FTP, HTTP, IMAP, LDAP, NTP, POP3, RTP, RTSP, SSH, SIP, SMTP, SNMP, Telnet, TFTP
Presentation	Data	Application		LPP, Kerberos, Radius
Session	Data			SSL — RPC, Sockets, L2TP
Transport	Segments	Transport		BGP — TCP, UDP
Network	Packets	Internet		ICMP, IGMP, OSPF, RIP — IPv4, IPv6 , IPSec
Data Link	Frames	Network Interface	LLC / MAC	ARP — IEEE 802.2, PPP — IEEE 802.3, 802.4, 802.5, 802.11
Physical	Bits			RJ45, DSL, SDH, 802.11, CAT 1-5, Coaxial Cable, FDDI, ATM, ISDN
Physical (Hardware)				

FIGURE 1.1
OSI and Internet Protocol layers.

communications across a network provides the path needed by applications above it, while it calls the next lower layer to send and receive packets that comprise the contents of that path. Two instances at the same layer are visualized as connected by a horizontal connection in that layer. Network architectures define the standards and techniques for designing and building communication systems for computers and other devices. To reduce the design complexity, most of the networks are organized as a series of layers or levels, each one builds upon the one below it. The basic idea of a layered architecture is to divide the design into small pieces. Each layer adds to the services provided by the lower layers in such a manner that the highest layer is provided a full set of services to manage communications and run the applications. The benefits of the layered models are modularity and clear interfaces, i.e., open architecture and comparability between the different providers' components.

1.1.2 Networking Services and Packet Encapsulation

In an *n*-layer architecture, layer *n* on one machine carries on a conversation with the layer *n* on another machine. The rules and conventions used in this conversation are collectively known as the layer-*n* protocol. The basic elements of a layered model are services, protocols, and interfaces. A service is a set of actions that a layer offers to another (higher) layer. A protocol is a set of rules that a layer uses to exchange information with a peer entity. These rules concern both the contents and the order of the messages used. Between the layers, service interfaces are defined. The data from one layer to another are sent through those interfaces.

Basically, a protocol is an agreement between the communicating parties on how communication is to proceed. Violating the protocol will make communication more difficult, if not impossible. As shown in Figure 1.1, several well-known protocols are presented at each layer. It is interesting to observe that the protocols' distribution is shaped like an hourglass, also referred to as a *narrow waist*, in which IP [139] occupies the most important position and it is the most critical protocol to support higher level protocols.

Between each pair of adjacent layers, there is an *interface*. The interface defines which primitive operations and services the lower layer offers to the upper layer adjacent to it. When the network designer decides how many layers to include in the network and what each layer should do, one of the main considerations is defining clean interfaces between adjacent layers. Doing so, in turn, requires that each layer should perform well-defined functions. In addition to minimizing the amount of information passed between layers, the neat interfaces also make it simpler to replace the implementation of one layer with a completely different implementation, because all that is required of new implementation is that it offers the same set of services to its upper layer neighbor as the old implementation (i.e., what a layer provides and how to use that service is more important than knowing how exactly it is implemented).

A set of layers and protocols is known as *network architecture*. The specification of architecture must contain enough information to allow an implementation to write the program or build the hardware for each layer so that it will correctly follow the appropriate protocol. Neither the details of implementation nor the specification of an interface is a part of network architecture because these are hidden away inside machines and not visible from outside. It is not even necessary that the interface on all machines in a network be the same, provided that each machine can correctly use all protocols. A list of protocols used by a certain system, one protocol per layer, is called *protocol stack*.

Figure 1.2 shows the TCP/IP layered protocol stack being used by a messaging application to send a message from left to right. In this example, the entities comprising the corresponding layers on different machines are called peers. In other words, it is the peers that communicate using protocols at its layer. Each layer passes data and control information to the layer immediately below it and a layer header is added to the data, which is called encapsulation, until the lowest layer is reached. Data at different encapsulation layers is

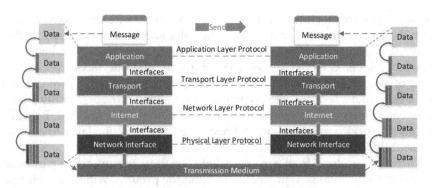

FIGURE 1.2
Service interfaces and packet encapsulation.

measured differently. Shown in Figure 1.1, at the physical layer, it is measured by bits; at the data link layer, it is measured by frames; at the network layer, it is measured by packets; and at the transport layers, it is measured by segments.

Below the network interface layer is the transmission medium, through which actual communication occurs. The receiver will process data in the reverse order as done by the sender, called decapsulation, to remove added headers at each layer, and finally reveal the message to the messaging application. The peer process abstraction is crucial to all network design. Using the layered approach, a basic principle is to ensure the independence of layers by defining services provided by each layer to the next higher layer without defining how the services are to be performed. This permits changes in a layer without affecting other layers.

1.2 Addresses

To understand how the TCP/IP protocol stack works, it is important to understand several addresses used at different protocol layers. In particular, they are: (a) data link layer: MAC addresses, (b) network layer: IP address, and (c) transport layer: port number.

In Figure 1.3, an email service example is presented to highlight the use of these addresses. MAC addresses allow networking devices to directly communicate with their intermediate neighboring devices at the data link layer. IP addresses are used for end-to-end addressing and multi-hop routing on the Internet. And finally, port numbers are used to identify which application on a host is to handle a received data packet. Altogether, these three important addresses are used to allow an application to deliver a message between two remote hosts through a hop-by-hop data forwarding approach through interconnected networking devices. We must note that through the data forwarding path, both IP addresses and port numbers will remain the same unless a Network Address Translation (NAT) service is used in between, and the Media Access Control (MAC) addresses will be changed on every en route.

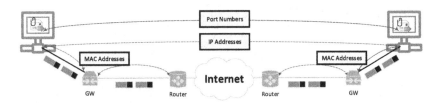

FIGURE 1.3
Addresses used in TCP/IP protocol stack.

1.2.1 MAC Address

Media Access Control (MAC) address of a device is a unique identifier assigned to network interface controllers for communications at the data link layer of a network segment. A MAC address format is presented in Figure 1.4. MAC addresses are used as the address for most IEEE 802 network technologies, including Ethernet and Wi-Fi. MAC addresses are most often assigned by the manufacturer of a Network Interface Controller (NIC) and are stored in its hardware, such as the network card's read-only memory or some other firmware mechanism. If assigned by the manufacturer, a MAC address usually encodes the manufacturer's registered identification number and may be referred to as the Burned-In Address (BIA). It may also be known as an Ethernet Hardware Address (EHA), Hardware Address or Physical Address. A network node may have multiple NICs and each NIC must have a unique MAC address. Sophisticated network equipment such as a multi-layer switch or router may require one or more permanently assigned MAC addresses. MAC addresses are formed according to the rules of one of three numbering name spaces managed by the Institute of Electrical and Electronics Engineers (IEEE): MAC-48, EUI-48, and EUI-64 [130].

Addresses can either be Universally Administered Addresses (UAA) or Locally Administered Addresses (LAA). A universally administered address is uniquely assigned to a device by its manufacturer. The first three octets identify the organization that issued the identifier and are known as the Organizationally Unique Identifier (OUI). The remainder of the address (three octets for MAC-48 and EUI-48 or five for EUI-64) are assigned by that organization in nearly any manner they please, subject to the constraint of uniqueness. A locally administered address is assigned to a device by a network administrator, overriding the burned-in address. For example, in a MAC

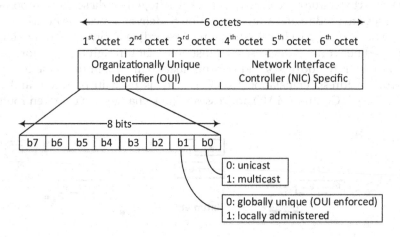

FIGURE 1.4
MAC address.

address 06-00-00-00-00-00, the first octet is 06 (hex), the binary form of which is 00000110, where the second-least-significant bit is 1, and thus it is a locally administered MAC address. MAC addresses are used for direct networking, where devices use MAC addresses within a local networking environment. A networking device needs to change its source and destination MAC addresses when forwarding it to the next network segment.

If the least significant bit of the first octet of an address is set to 0 (zero), the frame targets at one receiving NIC, in which this type of transmission is called unicast. If the least significant bit of the first octet is set to 1, the frame will still be sent only once; however, NICs will choose to accept it if they are allowed to receive multi-casting frames by using a configurable list of accepted multi-cast MAC addresses, and this MAC address is called multi-cast addressing. The IEEE has built in several special address types to allow more than one network interface card to be addressed at one time. These are examples of group addresses, as opposed to individual addresses; the least significant bit of the first octet of a MAC address distinguishes individual addresses from group addresses. Group addresses, like individual addresses, can be universally administered or locally administered. The group addresses are

- Packets sent to the broadcast address, all one bits, are received by all stations on a local area network. In hexadecimal, the broadcast address would be FF:FF:FF:FF:FF:FF. A broadcast frame is flooded and is forwarded to all other nodes.
- Packets sent to a multi-cast address are received by all stations on a LAN that have been configured to receive packets sent to that address.
- Functional addresses identify one or more Token Ring NICs that provide a particular service, defined in IEEE 802.5 [6].

1.2.2 IP Address (IPv4)

Every computer or device on the Internet must have a unique number assigned to it called the IP address. This IP address is used to recognize your particular computer out of the millions of other computers connected to the Internet.

In the mid-1990s, the Internet was a dramatically different network than when it was first established in the early 1980s. Today, the Internet has entered the public consciousness as the world's largest public data network, doubling in size every nine months. This is reflected in the tremendous popularity of the World Wide Web (WWW), the opportunities that businesses see in reaching customers from virtual storefronts, and the emergence of new types and methods of doing business. It is clear that expanding business and social awareness will continue to increase public demand for access to resources on the Internet. When IP was first standardized in September 1981, the specification required that each system attached to an IP-based Internet be assigned a unique, 32-bit

Internet address value. Some systems, such as routers which have interfaces to more than one network, must be assigned a unique IP address for each network interface. As shown below, IP addresses are categorized into 5 classes. This separation of classes of IPs results in what is known as classful IP addresses.

Class A
Start IP: 0. 0. 0. 0   → 00000000.00000000.00000000.00000000
End IP: 127.255.255.255   → 01111111.11111111.11111111.11111111

Class B
Start IP: 128. 0. 0. 0   → 10000000.00000000.00000000.00000000
End IP: 191.255.255.255   → 10111111.11111111.11111111.11111111

Class C
Start IP: 192. 0. 0. 0   → 11000000.00000000.00000000.00000000
End IP: 239.255.255.255   → 11011111.11111111.11111111.11111111

Class D
Start IP: 224. 0. 0. 0   → 11100000.00000000.00000000.00000000
End IP: 239.255.255.255   → 11101111.11111111.11111111.11111111

Class E
Start IP: 240. 0. 0. 0   → 11110000.00000000.00000000.00000000
End IP: 255.255.255.255   → 11111111.11111111.11111111.11111111

The first part of the IP address, underlined in the snippet above identifies the network on which the host resides, while the second part identifies the particular host on the given network. This created the two-level addressing hierarchy.

In 1985, RFC 950 defined a standard procedure to support the subnetting, or division, which was introduced to overcome some of the problems that parts of the Internet were beginning to experience with the classful two-level addressing hierarchy. In 1987, RFC 1009 specified how a subnetted network could use more than one subnet mask. When an IP network is assigned more than one subnet mask, it is considered a network with "Variable Length Subnet Masks (VLSM)" since the extended-network-prefixes have different lengths.

1.2.2.1 Classless Inter-Domain Routing

By 1992, the exponential growth of the Internet was beginning to raise serious concerns among members of the IETF about the ability of Internet's routing system to scale and support future growth. These problems were related to:

- The near-term exhaustion of the Class B network address space.
- The rapid growth in the size of the global Internet's routing tables.
- The eventual exhaustion of the 32-bit IPv4 address space.

Projected Internet growth figures made it clear that the first two problems were likely to become critical by 1994 or 1995. The response to these immediate challenges was the development of the concept of Supernetting or Classless Inter-Domain Routing (CIDR). The third problem, which is of a more long-term nature, is currently being explored by the IP Next Generation (IPng or IPv6) working group of the IETF.

CIDR was officially documented in September 1993 in RFC 1517, 1518, 1519, and 1520. CIDR supports two important features that benefit the global Internet routing system, each of which is detailed next.

CIDR Promotes the Efficient Allocation of the IPv4 Address Space

CIDR eliminates the traditional concept of Class A, Class B, and Class C network addresses with the generalized concept of a "network-prefix." By having routers use the network-prefix to determine the dividing point between the network address and the host address, CIDR supports the deployment of *arbitrarily sized* networks rather than the standard 8-bit, 16-bit, or 24-bit network numbers associated with classful addressing. This enables the efficient allocation of the IPv4 address space which will allow the continued growth of the Internet until IPv6 is deployed.

In the CIDR model, each piece of routing information is advertised with a bit mask (or prefix-length). The prefix-length is a way of specifying the number of leftmost contiguous bits in the network-portion of each routing table entry. For example, a network with 20 bits of network-number and 12-bits of host-number would be advertised with a 20-bit prefix length (a /20). The clever thing is that the IP address advertised with the /20 prefix could be a former Class A, Class B, or Class C. Routers that support CIDR do *not* make assumptions based on the first 3-bits of the address, they rely on the prefix-length information provided with the route.

In a classless environment, prefixes are viewed as bitwise contiguous blocks of the IP address space. For example, all prefixes with a /20 prefix represent the same amount of address space (2^{12} or 4,096 host addresses). Furthermore, a /20 prefix can be assigned to a traditional Class A, Class B, or Class C network number. The snippet below shows how each of the following /20 blocks represents 4,096 host addresses - 10.23.64.0/20, 130.5.0.0/20, and 200.7.128.0/20.

Traditional A: 10.23.64.0/20 → <u>00001010.00010111.0100</u>0000.00000000

Traditional B: 130.5.0.0/20 → <u>10000010.00000101.0000</u>0000.00000000

Traditional C: 200.7.128.0/20 → <u>11001000.00000111.1000</u>0000.00000000

Table 1.1 provides information about the most commonly deployed CIDR address blocks. Referring to the table, you can see that a /15 allocation can also be specified using the traditional dotted-decimal mask notation of 255.254.0.0. Also, a /15 allocation contains a bitwise contiguous block of

TABLE 1.1

CIDR Address Blocks

CIDR prefix-length	Dotted decimal	# of Individual Addresses	# of Classful Networks
/13	255.248.0.0	512K	8Bs or 2048Cs
/14	255.252.0.0	256K	4Bs or 1024Cs
/15	255.254.0.0	128K	2Bs or 512Cs
/16	255.255.0.0	64K	1B or 256Cs
/17	255.255.128.0	32K	128Cs
/18	255.255.192.0	16K	64Cs
/19	255.255.224.0	8K	32Cs
/20	255.255.240.0	4K	16Cs
/21	255.255.248.0	2K	8Cs
/22	255.255.252.0	1K	4Cs
/23	255.255.254.0	512	2Cs
/24	255.255.255.0	256	1C
/25	255.255.255.128	128	1/2C
/26	255.255.255.192	64	1/4C
/27	255.255.255.224	32	1/8C

128K (131,072) IP addresses which can be classfully interpreted as 2 Class B networks or 512 Class C networks.

How does all of this lead to the efficient allocation of the IPv4 address space? In a classful environment, an Internet Service Provider (ISP) can only allocate /8, /16, or /24 addresses. In a CIDR environment, the ISP can carve out a block of its registered address space that specifically meets the needs of each client, provides additional room for growth, and does not waste a scarce resource.

Assume that an ISP has been assigned the address block 206.0.64.0/18. This block represents 16,384 (2^{14}) IP addresses which can be interpreted as 64 /24s. If a client requires 800 host addresses, rather than assigning a Class B (and wasting 64,700 addresses) or four individual Class Cs (and introducing 4 new routes into the global Internet routing tables), the ISP could assign the client the address block 206.0.68.0/22, a block of 1,024 (2^{10}) IP addresses (4 contiguous /24s). The efficiency of this allocation is illustrated as follows:

ISP's Block:	206.0.64.0/18	→ <u>111001110.00000000.01</u>000000.00000000
Client Block:	206.0.68.0/22	→ <u>111001110.00000000.010001</u>00.00000000
Class C #0:	206.0.68.0/24	→ <u>111001110.00000000.01000100</u>.00000000
Class C #1:	206.0.69.0/24	→ <u>111001110.00000000.01000101</u>.00000000
Class C #2:	206.0.70.0/24	→ <u>111001110.00000000.01000110</u>.00000000
Class C #3:	206.0.71.0/24	→ <u>111001110.00000000.01000111</u>.00000000

CIDR Supports Route Aggregation

CIDR supports route aggregation where a single routing table entry can represent the address space of perhaps thousands of traditional classful routes. This allows a single routing table entry to specify how to route traffic to many individual network addresses. Route aggregation helps control the amount of routing information in the Internet's backbone routers, reduces route flapping (rapid changes in route availability), and eases the local administrative burden of updating external routing information. Without the rapid deployment of CIDR in 1994 and 1995, the Internet routing tables would have in excess of 70,000 routes (instead of the current 30,000+) and the Internet would probably not be functioning today!

Host Implications for CIDR Deployment

It is important to note that there may be severe host implications when you deploy CIDR-based networks. Since many legacy hosts are classful, their user interface will not permit them to be configured with a mask that is shorter than the "natural" mask for a traditional classful address. For example, potential problems could exist if you wanted to deploy 200.25.16.0 as a /20 to define a network capable of supporting 4,094 (212-2) hosts. The software executing on each end station might not allow a traditional Class C (200.25.16.0) to be configured with a 20-bit mask since the natural mask for a Class C network is a 24-bit mask. If the host software supports CIDR, it will permit shorter masks to be configured.

However, there will be no host problems if you were to deploy the 200.25.16.0/20 (a traditional Class C) allocation as a block of 16 /24s since non-CIDR hosts will interpret their local /24 as a Class C. Likewise, 130.14.0.0/16 (a traditional Class B) could be deployed as a block of 255 /24s since the hosts will interpret the /24s as subnets of a /16. If host software supports the configuration of shorter than expected masks, the network manager has tremendous flexibility in network design and address allocation.

CIDR Address Allocation Example

For this example, assume that an ISP owns the address block 200.25.0.0/16. This block represents 65,536 (2^{16}) IP addresses (or 256 /24s). From the 200.25.0.0/16 block, it wants to allocate the 200.25.16.0/20 address sub-blocks. This smaller block represents 4,096 (2^{12}) IP addresses (or 16 /24s).

Address Block: 200.25.16.0/20 → 11001000.00011001.00010000.00000000
In a classful environment, the ISP is forced to use the /20 as 16 individual /24s as below:

Network # 0:200.25.16.0/24 → 11001000.00011001.0001**0001**.00000000
Network # 1:200.25.17.0/24 → 11001000.00011001.0001**0001**.00000000
Network # 2:200.25.18.0/24 → 11001000.00011001.0001**0010**.00000000

Network # 3:200.25.19.0/24   → 11001000.00011001.0001**0011**.00000000
Network # 4:200.25.20.0/24   → 11001000.00011001.0001**0100**.00000000

.
.

Network # 13:200.25.29.0/24   → 11001000.00011001.0001**1101**.00000000
Network # 14:200.25.30.0/24   → 11001000.00011001.0001**1110**.00000000
Network # 15:200.25.31.0/24   → 11001000.00011001.0001**1111**.00000000

If you look at the ISP's /20 address block as a pie, in a classful environment it can only be cut into 16 equal-size pieces. This is illustrated in Figure 1.5.

However, in a classless environment, the ISP is free to cut up the pie any way it wants. It could slice up the original pie into 2 pieces (each 1/2 of the address space) and assign one portion to Organization A, then cut the other half into 2 pieces (each 1/4 of the address space) and assign one piece to Organization B, and finally slice the remaining fourth into 2 pieces (each 1/8 of the address space) and assign it to Organization C and Organization D. Each of the individual organizations is free to allocate the address space within its "Intranetwork" as it sees fit. The steps followed are shown below. The slicing of the IP space, in this case, is illustrated in Figure 1.6.

Step #1 Divide the address block 200.25.16.0/20 into two equal size slices. Each block represents one-half of the address space or 2,048 (2^{11}) IP addresses.

ISP's Block: 200.25.16.0/20   → 11001000.00011001.00010000.00000000
Org A: 200.25.16.0/21   → 11001000.00011001.000100**00**.00000000
Reserved: 200.25.24.0/21   → 11001000.00011001.000110**00**.00000000

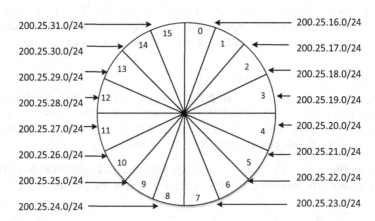

FIGURE 1.5
Slicing the pie – classful environment.

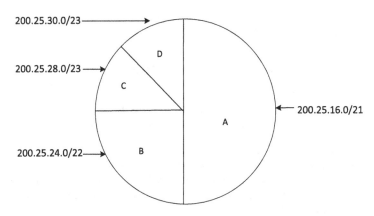

FIGURE 1.6
Slicing the pie – CIDR environment.

Step #2 Divide the reserved block (200.25.24.0/21) into two equal size slices. Each block represents one-fourth of the address space or 1,024 (2^{10}) IP addresses.

Reserved: 200.25.24.0/21 → 11001000.00011001.00011000.00000000
Org B: 200.25.24.0/22 → 11001000.00011001.00011000.00000000
Reserved: 200.25.28.0/22 → 11001000.00011001.00011100.00000000

Step #3 Divide the reserved address block (200.25.28.0/22) into two equal size blocks. Each block represents one-eighth of the address space or 512 (2^{9}) IP addresses.

Reserved: 200.25.28.0/22 → 11001000.00011001.00011100.00000000
Org C: 200.25.28.0/23 → 11001000.00011001.00011100.00000000
Org D: 200.25.30.0/23 → 11001000.00011001.00011110.00000000

1.2.2.2 Private IPs

With CIDR, we can assign 32 bits freely to describe the network address portion and host address portion. However, the Internet is still facing an address shortage issue. To further deal with the address shortage issue, RFC 1918 requests that organizations make use of the private Internet address space for hosts that require IP connectivity within their enterprise network, but do not require external connections to the global Internet. For this purpose, the

Internet Assigned Numbers Authority (IANA) has reserved the following three address blocks for private networks:

10.0.0.0 – 10.255.255.255(10.0.0.0/8)
172.16.0.0 – 172.31.255.255(172.16.0.0/12)
192.168.0.0 – 192.168.255.255(192.168.0.0/16)

Any organization that elects to use addresses from these reserved blocks can do so without contacting the IANA or an Internet registry. Since these addresses are never injected into the global Internet routing system, the address space can simultaneously be used by many different organizations. The disadvantage to this addressing scheme is that it requires an organization to use a Network Address Translator (NAT) for local addresses to have global Internet access. However, the use of the private address space and a NAT make it much easier for clients to change their ISP without the need to renumber or "punch holes" in a previously aggregated advertisement. The benefits of this addressing scheme to the Internet is that it reduces the demand for IP addresses so that large organizations may require only a small block of the globally unique IPv4 address space.

1.2.3 IP Address (IPv6)

IPv6 is designed to address the IP address shortage issue of IPv4 due to its limited 32-bit address space. IPv6 addresses have 128 bits. To put that number in perspective, the number of atoms on the surface of the earth is 1.26×10^{34}, and the total number of IPv6 addresses is 3.4×10^{38}. Thus, we could have an IP address for every atom on the surface of the earth and have enough left over to repeat the process a hundred times. The IPv6 address space implements a very different design philosophy than in IPv4, in which subnetting was used to improve the efficiency of utilization of the small address space.

Since in IPv6 the address space is deemed large enough, a local area subnet always uses 64 bits for the host portion of the address, designated as the interface identifier, while the most significant 64 bits are used as the routing prefix. The identifier is only unique within the subnet to which a host is connected. IPv6 has a mechanism for automatic address detection, so that address autoconfiguration always produces unique assignments.

1.2.3.1 Address Representation

The 128 bits of an IPv6 address is represented in 8 groups of 16 bits each. Each group is written as four hexadecimal digits (sometimes called hextets) and the groups are separated by colons (:). An example of this representation is 2001:0db8:0000:0000:0000:ff00:0042:8329. For convenience, an IPv6 address may be abbreviated to shorter notations by application of the following rules:

1. One or more leading zeroes from any groups of hexadecimal digits are removed; this is usually done to either all or none of the leading zeroes. For example, the group 0042 is converted to 42.

2. Consecutive sections of zeroes are replaced with a double colon (::).

3. The double colon may only be used once in an address, as multiple uses would render the address indeterminate. RFC 5952 recommends that a double colon not be used to denote an omitted single section of zeroes.

As an example of an application of these rules, consider the initial address to be 2001:0db8:0000:0000:0000:ff00:0042:8329. After removing all leading zeroes in each group, we have the address 2001:db8:0:0:0:ff00:42:8329. Next, after omitting consecutive sections of zeroes, we get 2001:db8::ff00:42:8329. Similarly, the loopback address, 0000:0000:0000:0000:0000:0000:0000:0001, may be abbreviated to ::1 by using the abbreviation rule. As an IPv6 address may have more than one representation, the IETF has issued a proposed standard for representing them in text [146].

1.2.3.2 Address Uniqueness

Hosts verify the uniqueness of addresses assigned by sending a neighbor solicitation message asking for the Link Layer address of the IP address. If any other host is using that address, it responds. However, MAC addresses are designed to be unique on each network card which minimizes chances of duplication.

The host first determines if the network is connected to any routers at all, because if not, then all nodes are reachable using the link-local address that already is assigned to the host. The host will send out a Router Solicitation message to the all-routers multi-cast group with its link-local address as the source. If there is no answer after a predetermined number of attempts, the host concludes that no routers are connected. If it does get a response from a router, there will be network information inside that is needed to create a globally unique address. There are also two flag bits that tell the host whether it should use Dynamic Host Configuration Protocol (DHCP) to get further information and addresses:

1. The *Managed* bit, which indicates whether or not the host should use DHCP to obtain additional addresses.

2. The *Other* bit, which indicates whether or not the host should obtain other information through DHCP.

The other information consists of one or more prefix information options for the subnets that the host is attached to, a lifetime for the prefix, and two flags:

- *On-link*: If this flag is set, the host will treat all addresses on the specific subnet as being on-link, and send packets directly to them instead of sending them to a router for the duration of the given lifetime.
- *Address*: This is the flag that tells the host to actually create a global address.

1.2.3.3 Link-local Address

All interfaces of IPv6 hosts require a link-local address. A link-local address is derived from the MAC address of the interface and the prefix fe80::/10. The process involves filling the address space with prefix bits left-justified to the most-significant bit and filling the MAC address in EUI-64 format into the least-significant bits. If any bits remain to be filled between the two parts, those are set to zero.

The uniqueness of the address on the subnet is tested with the Duplicate Address Detection (DAD) method.

1.2.3.4 Global Addressing

The assignment procedure for global addresses is similar to local address construction. The prefix is supplied from router advertisements on the network. Multiple prefix announcements cause multiple addresses to be configured.

Stateless Address Autoconfiguration (SLAAC) requires a /64 address block, as defined in RFC 4291. Local Internet registries are assigned at least /32 blocks, which they divide among subordinate networks. The initial recommendation stated assignment of a /48 subnet to end-consumer sites (RFC 3177). This was replaced by RFC 6177, which recommends giving home sites significantly more than a single /64 but does not recommend that every home site is given a /48 either. /56s are specifically considered. It remains to be seen if ISPs will honor this recommendation. For example, during initial trials, Comcast customers were given a single /64 network.

IPv6 addresses are classified by three types of networking methodologies:

- unicast addresses identify each network interface,
- anycast addresses identify a group of interfaces, usually at different locations of which the nearest one is automatically selected, and
- multi-cast addresses are used to deliver one packet to many interfaces.

The broadcast method is not implemented in IPv6. Each IPv6 address has a scope, which specifies in which part of the network it is valid and unique.

Some addresses are unique only on the local (sub-)network. Others are globally unique.

Some IPv6 addresses are reserved for special purposes, such as loopback, 6-to-4 tunneling, and Teredo tunneling, as outlined in RFC 5156. Also, some address ranges are considered special, such as link-local addresses for use on the local link only, Unique Local addresses (ULA), as described in RFC 4193, and solicited-node multi-cast addresses used in the Neighbor Discovery Protocol.

1.2.4 Port Number

When information is sent over the Internet to a computer, it accepts that information by using Transmission Control Protocol (TCP) or User Datagram Protocol (UDP) ports. When a program on the computer sends or receives data over the Internet, it sends that data to an IP specific port on the remote computer and receives the data on a usual address and a sly random port on its own computer. If it uses the TCP protocol to send and receive the data, then it will connect and bind itself to a TCP port. If it uses the UDP protocol to send and receive data, it will use a UDP port. In order for a web server to accept connections from remote computers, such as yourself, it must bind the web server application to a local port. It will then use this port to listen for and accept connections from remote computers. Web servers typically bind to the TCP port 80, which is what the Hypertext Transfer Protocol (HTTP) uses by default, and then will wait and listen for connections from remote devices. Once a device is connected, it will send the requested web pages to the remote device, and when complete, disconnect the connection. On the other hand, if a remote user connects to a web server, it would work in reverse. Your web browser would pick a random TCP port from a certain range of port numbers and attempt to connect to port 80 on the IP address of the web server. When the connection is established, the web browser will send the request for a particular web page and receive it from the web server. Then both computers will disconnect the connection. Thus, we can view a port number as the address for an application.

In TCP and UDP networks, a port is an endpoint to a logical connection and the way a client program specifies a specific server program on a computer in a network. The port number identifies what type of port it is. For example, port 80 is usually used for HTTP traffic. Some ports have numbers that are assigned to them by the IANA, and these are called the "well-known ports" which are specified in RFC 1700, and there is a total $2^{16} = 65536$ port numbers available for each host, for each protocol.

Using TCP, the host sending the data connects directly to the computer it is sending the data to and stays connected for the duration of the transfer. With this method, the two communicating hosts can guarantee that the data has arrived safely and correctly, and then they disconnect the connection. This method of transferring data tends to be quicker and more reliable but puts a higher load on the computer as it has to monitor the connection and the data going across it.

Using UDP on the other hand, the information is packaged into a nice little package and releases it into the network with the hopes that it will get to the right place. What this means is that UDP does not connect directly to the receiving computer like TCP does, but rather sends the data out and relies on the devices in between the sending computer and the receiving computer to get the data where it is supposed to go properly. This method of transmission does not provide any guarantee that the data you send will ever reach its destination. On the other hand, this method of transmission has a very low overhead and is therefore very popular to use for services that are not that important to work on the first try.

1.3 Physical, Logical, and Overlay Networks

The layered structure of computer networks provides an efficient and effective approach to describe complicated dependencies among different networking services. However, the layered structure also blocks access to some network functions on lower layers. Virtual networking solutions are a group of techniques that enable access to network functions at any layer. To better illustrate what virtual networks are, we first need to understand differences among several network terms: physical networks, logical (virtual) networks, and overlay networks.

1.3.1 Physical Networks

A physical network is visible and physically presented to connect physical computers as shown in Figure 1.7(a). We can go to any physical server (or desktop PC for that matter) and check the status of the network connection by seeing the "media state"; i.e., if it is enabled. We can check its speed, how long it has been up, and what its connectivity state is as defined by the operating system on that server. We are comfortable with network monitoring tools and agents for physical servers and their physical network connections.

1.3.2 Logical Networks

A logical network is a virtual representation of a network that appears to the user as an entirely separate and self-contained network even though it might physically be only a portion of a larger network or a local area network. It might also be an entity that has been created out of multiple separate networks and made to appear as a single network. This is often used in virtual environments where there are physical and virtual networks running together; so, out

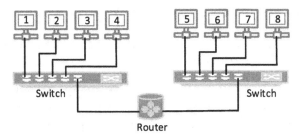

(*a*) Physical Network.

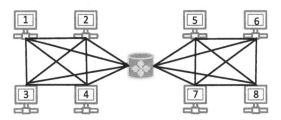

(*b*) Logical Network (Link-layer View).

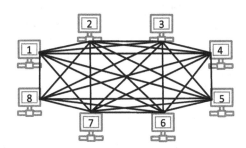

(*c*) Logical Network (Network-layer View).

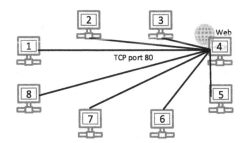

(*d*) Logical Network (Application-layer View-Web Service).

FIGURE 1.7
Physical and logical view of networks.

of convenience and function, separate networks can be made into a single logical network.

A logical network is usually represented based on which level to view it. As shown in Figure 1.7(b)-(d), a logical network may be viewed differently at different protocol layers. At the link layer, there are two isolated and fully meshed link-layer networks, which are isolated by a network-layer router; at the network layer, if the routing is set up correctly, it will be represented as a fully meshed peer-to-peer network. At the application level, e.g., a web service model, the logical network can be viewed as a one-to-many client-service based networking service model.

In Figure 1.7(c)-(d), we can see that a logical network, unlike a physical network, often spans multiple physical devices such as network nodes and networking equipment that are often parts of separate physical networks. For example, a logical network can be made up of elements from separate networks with devices located around the globe, as in a global enterprise where the computers of site managers from different countries might be connected as a single logical network in order to foster quick and hassle-free communication even though they are physically separated by continents.

This concept is very important for distributed applications as it binds the distributed components, more or less, as a single group or a single entity. In this way, logical components can be arranged into groups that represent business environments or departments such as finance, engineering, human resource or quality assurance. Those environments are then treated as a single logical network even though their physical components might be located in different geographical zones.

1.3.3 Overlay Networks

An overlay network is a virtual network of nodes and logical links that are built on top of an existing network with the purpose of implementing a network service that is not available in the existing network. That overlay networks implement a service or application is the subtle difference between them and logical networks. For example, Napster is a peer-to-peer overlay network for music file sharing service. Nodes in the overlay network can be thought of as being connected by virtual links, each of which corresponds to a path in the underlying network, perhaps through many physical links. Distributed systems such as peer-to-peer networks and client-server applications are overlay networks because their nodes run on top of the Internet. Interestingly, while the Internet was originally built as an overlay upon the telephone network, today (through the advent of Voice-over-IP (VoIP)), the telephone network is increasingly turning into an overlay network built on top of the Internet.

Similar to logical networks, overlay networks run as independent virtual networks on top of a physical network infrastructure. These virtual network overlays allow service providers to provision and orchestrate networks alongside other virtual resources. They also offer a new path to converge networks and programmability. Using a network overlay is one way to implement a SDN architecture. In the context of SDNs, an overlay network uses virtual links to connect to the underlying physical network (e.g., Ethernet switches or routers). As a convention, we usually use virtual networks instead of using overlay networks in the context of SDN technologies.

1.4 Computer Networking Services

1.4.1 Address Resolution Protocol

The Address Resolution Protocol (ARP) is a communication protocol used for discovering the link layer address, such as a MAC address, associated with a given network layer address, typically an IPv4 address. This mapping is a critical function in the Internet protocol suite. ARP was defined in RFC 826.

ARP uses a simple message format containing one address resolution request or response, which is shown in Figure 1.8. The size of the ARP message depends on the upper layer and lower layer address sizes, which are given by the type of networking protocol (usually IPv4) in use and the type of hardware or virtual link layer that the upper layer protocol is running on. The message header specifies these types, as well as the size of addresses of each. The message header is completed with the operation code for request (1) and reply (2). The payload of the packet consists of four addresses, the hardware and protocol address of the sender and receiver hosts.

As shown in Figure 1.8, when a IPv4 network running on Ethernet, the packet has 48-bit fields for the Sender Hardware Address (SHA) and Target Hardware Address (THA), and 32-bit fields for the corresponding sender

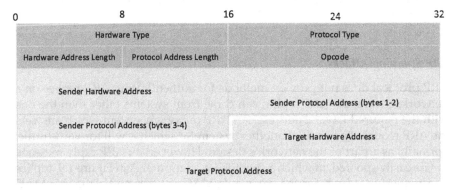

FIGURE 1.8
ARP protocol packet format.

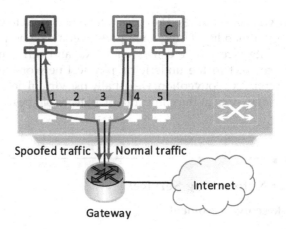

FIGURE 1.9
ARP and ARP spoofing attack.

and target protocol addresses (SPA and TPA). Thus, the ARP packet size in this case is 28 bytes. The EtherType for ARP is 0x0806.

In Figure 1.9, it shows two computers B and C are connected to each other in a switch, with no intervening gateways or routers. B has a packet to send to C. Through DNS, it determines that C has the IP address 192.168.0.55. To send the message, it also requires C's MAC address. First, B uses a cached ARP table to look up 192.168.0.55 for any existing records of C's MAC address (00:eb:24: b2:05:ac). If the MAC address is found, it sends an Ethernet frame with destination address 00:eb:24:b2:05:ac, containing the IP packet onto the link. If the cache did not produce a result for 192.168.0.55, B has to send a broadcast ARP message (destination FF:FF:FF:FF:FF:FF MAC address), which is accepted by all computers, requesting an answer for 192.168.0.55. C responds with its MAC and IP addresses. C may insert an entry for B into its ARP table for future use. B then caches the response information in its ARP table and can now send the packet to C.

ARP Spoofing Attacks

ARP protocol does not provide methods for authenticating ARP replies on a network; as a result, ARP replies can come from systems other than the one with the required Layer 2 address. An ARP proxy is a system which answers the ARP request on behalf of another system for which it will forward traffic, normally as a part of the network's design. However, an ARP reply message can be easily spoofed, in which a spoofer, i.e., computer A in Figure 1.9, replies to a request sent by B for the gateway's MAC address and claim A's MAC address is the Gateway's MAC address. After receiving the ARP reply message, B will put A's MAC address as the destination MAC address in the packet, and thus A can perform a Man-in-The-Middle (MITM) or Denial-of-Service (DoS) attack to computer B.

1.4.2 Dynamic Host Configuration Protocol

Dynamic Host Configuration Protocol (DHCP) is a network management protocol used to dynamically assign an Internet Protocol (IP) address to any device, or node, on a network so they can communicate using IP. DHCP automates and centrally manages these configurations rather than requiring network administrators to manually assign IP addresses to all network devices. DHCP assigns new IP addresses in each location when devices are moved from place to place, which means network administrators do not have to manually initially configure each device with a valid IP address or reconfigure the device with a new IP address if it moves to a new location on the network. Versions of DHCP are available for use in Internet Protocol version 4 (IPv4) and Internet Protocol version 6 (IPv6).

The DHCP employs a connectionless service model, using the User Datagram Protocol (UDP). It is implemented with two UDP port numbers for its operations, which are the same as for the Bootstrap Protocol (BOOTP). UDP port number 67 is the destination port of a server, and UDP port number 68 is used by the client. DHCP dynamically assigns IP addresses to DHCP clients and to allocate TCP/IP configuration information to DHCP clients. This includes *IP address* and *subnet mask* information, *default gateway IP addresses* and *domain name system (DNS) addresses.*

DHCP is a client-server protocol in which servers manage a pool of unique IP addresses, as well as information about client configuration parameters, and assign addresses out of those address pools. DHCP-enabled clients send a request to the DHCP server whenever they connect to a network.

Clients configured with DHCP broadcast a request to the DHCP server and request network configuration information for the local network to which they are attached. A client typically broadcasts a query for this information immediately after booting up. The DHCP server responds to the client request by providing IP configuration information previously specified by a network administrator. This includes a specific IP address as well as for the time period, also called a lease, for which the allocation is valid. When refreshing an assignment, a DHCP client requests the same parameters, but the DHCP server may assign a new IP address based on policies set by administrators.

A DHCP server manages a record of all the IP addresses it allocates to network nodes. If a node is relocated in the network, the server identifies it using its MAC address, which prevents accidentally configuring multiple devices with the same IP address.

DHCP is limited to a specific Local Area Network (LAN), which means a single DHCP server per LAN is adequate, or two servers for use in case of a fail-over. Larger networks may have a Wide Area Network (WAN) containing multiple individual locations. Depending on the connections between these points and the number of clients in each location, multiple DHCP servers can be set up to handle the distribution of addresses. If network administrators want a DHCP server to provide addressing to multiple subnets on a given network, they must configure DHCP relay services located on interconnecting routers that DHCP requests have to cross. These agents relay messages between DHCP clients and servers located on different subnets.

DHCP lacks any built-in mechanism that would allow clients and servers to authenticate each other. Both are vulnerable to deception (one computer pretending to be another) and to attack, where rogue clients can exhaust a DHCP server's IP address pool.

1.4.3 Domain Name System

The Domain Name System (DNS) is a hierarchical decentralized naming system for computers, services, or other resources connected to the Internet or a private network. It associates various information with domain names assigned to each of the participating entities. Most prominently, it translates more readily memorized domain names to the numerical IP addresses needed for locating and identifying computer services and devices with the underlying network protocols. By providing a worldwide, distributed directory service, DNS is an essential component of the functionality on the Internet that has been in use since 1985.

The DNS delegates the responsibility of assigning domain names and mapping those names to Internet resources by designating authoritative name servers for each domain. Network administrators may delegate authority over sub-domains of their allocated name space to other name servers. This mechanism provides distributed and fault-tolerant service and was designed to avoid a single large central database.

The DNS also specifies the technical functionality of the database service that is at its core. It defines the DNS protocol, a detailed specification of the data structures and data communication exchanges used in the DNS, as part of the Internet Protocol Suite. Historically, other directory services preceding DNS were not scalable to large or global directories as they were originally based on text files, prominently the host's file.

The Internet maintains two principal namespaces, the domain name hierarchy and the Internet Protocol (IP) address spaces. The DNS maintains the domain name hierarchy and provides translation services between it and the address spaces. Internet name servers and a communication protocol implement the DNS. A DNS name server is a server that stores the DNS records for a domain; a DNS name server responds with answers to queries against its database.

The most common types of records stored in the DNS database are for Start of Authority (SOA), IP addresses (A and AAAA), SMTP mail exchangers (MX), name servers (NS), pointers for reverse DNS lookups (PTR), and domain name aliases (CNAME). Although not intended to be a general-purpose database, DNS can store records for other types of data for either automatic lookups, such as DNS Security (DNSSEC) records, or for human queries such as responsible person (RP) records. As a general-purpose database, the DNS has also been used in combating unsolicited email (spam) by storing a real-time blackhole list. The DNS database is traditionally stored in a structured zone file.

1.4.4 Network Address Translation

1.4.4.1 What is NAT

Network Address Translation (NAT) is the process where a network device, usually a firewall, assigns a public address to a computer (or group of computers) inside a private network. The main use of NAT is to limit the number of public IP addresses an organization or company must use, for both economy and security purposes.

The common form of network translation involves a large private network using private IP addresses in any of the following given ranges:

- 10.0.0.0 to 10.255.255.255 (10.0.0.0/8)
- 172.16.0.0 to 172.31.255.255 (172.16.0.0/12), or
- 192.168.0 0 to 192.168.255.255 (192.168.0.0/16)

The private addressing scheme works well for computers that only have to access resources inside the network, like workstations needing access to file servers and printers. Routers inside the private network can route traffic between private addresses with no trouble. However, to access resources outside the network, like the Internet, these computers have to have a public address in order for responses to their requests to return to them.

For a stricter definition of NAT, it means a one-to-one, one-to-many, or many-to-many private-to-public IP addresses translation or it is represented as k-to-n translation and usually $k \geq n$, which allows one or multiple public IP address(es) to represent one or multiple internal private IP address(es). When $k > n$, multiple private IP addresses need to share a small number of public IP addresses, and this can deduce to another public IP shortage issue since all the private IP addresses need to compete to get a public IP to access to the public domain. In order to address this issue, we can use both IP addresses and port numbers, e.g., TCP or UDP ports, to use both a public IP address a port number to uniquely translate a private IP address to a public IP address and a port number based on a small number of public IP addresses. This approach is called Network and Port Address Translation (NPAT), or simple Port Address Translation (PAT). Due to the tradition and convenience, we do not strictly differentiate between NAT, NPAT, or PAT, and call them collectively NAT for simplicity. It usually means that the translation includes both IP and port addresses without a special notice, otherwise.

Usually, NAT is transparent to end users. A workstation inside a network makes a request to a computer on the Internet. Routers within the network recognize that the request is not for a resource inside the network, so they send the request to the firewall. The firewall sees the request from the computer with the internal private IP address. It then makes the same request to the Internet using its own public IP address, and returns the response from the Internet resource to the computer inside the private network. From the perspective of the resource on the Internet, it is sending information to the address of the firewall. From the perspective of the workstation, it appears that communication is directly with the destination site on the Internet. When NAT is used in this way, all users inside the private network access the Internet have the same public IP address when they use the Internet. That means only one public address is needed for hundreds or even thousands of users within the private network domain.

Most modern firewalls are stateful; that is, they are able to set up the connection between the internal workstation and the Internet resource. They can keep on tracking the details of the connection, like ports, packet order, and the IP addresses involved, which means "stateful." In this way, they are able to keep track of the session composed of communication between the workstation and the firewall, and the firewall with the Internet. When the session ends, the firewall discards all of the information about the connection.

There are other uses for NAT beyond simply allowing workstations with internal IP addresses to access the Internet. In large networks, some servers may act as Web servers and require access from the Internet. These servers are assigned public IP addresses on the firewall, allowing the public to access the servers only through that IP address. However, as an additional layer of security, the firewall acts as the intermediary between the outside world and the protected internal network. Additional rules can be added, including which ports can be accessed at that IP address. Using NAT in this way allows network and security engineers to more efficiently route internal network traffic to the same resources, and allow access to more ports, while restricting the access at the firewall. It also allows detailed logging of communications between the network and the outside world.

Additionally, NAT can be used to allow selective access to the outside of the network, too. Workstations or other computers requiring special access outside the network can be assigned specific external public IP address using NAT, allowing them to communicate with computers and applications that require a unique public IP address. Again, the firewall acts as the intermediary, and can control the session in both directions, restricting port access and protocols.

1.4.4.2 PREROUTING and POSTROUTING

In NAT table, it contains two chains: PREROUTING and POSTROUTING. The PREROUTING is done at first when a packet is received and is thus routed based on where it is going (destination). After all other routing rules have been applied, the POSTROUTING chain will determine where it goes based on where it came from (source).

For example, on a server, incoming ports that are to be forwarded (NATed) are all defined in the PREROUTING chain as DNAT, and all packets that come from the NATed interfaces, go through the POSTROUTING chain as SNAT, and consequently (in this case), go through the filter FORWARD chain. As the name suggested, PREROUTING means the NAT-based IP/port changes need to have occurred before routing. This will allow an incoming packet to change its destination address (i.e., use DNAT in iptables setup) to a NATed IP address (e.g., a private IP), and then the routing will decide how to forward to the internal interface-based on the changed destination IP address. On the contrary, POSTROUTING chain will be evaluated after the routing decision, i.e., change the source IP address to a public IP address associated to an outgoing interface.

1.4.4.3 Netfilter and NAT

The basic firewall software most commonly used in Linux is called iptables. The iptables' firewall works by interacting with the packet filtering hooks in the Linux kernel's networking stack. These kernel hooks are known as the netfilter framework. Every packet that enters a networking system (incoming or outgoing) will trigger these hooks as it progresses through the stack, allowing programs that register with these hooks to interact with the traffic at key points. The kernel modules associated with iptables register at these hooks in order to ensure that the traffic conforms to the conditions laid out by the firewall rules.

The NAT packet processing procedure is highlighted in Figure 1.10. The PREROUTING chain is responsible for packets that just arrived at the network interface before routing decision has taken place. As a result, it is not yet known whether the packet would be interpreted locally or whether it would be forwarded to another machine located at another network interface. After the packet has passed the PREROUTING chain, e.g., it may change the packet's public destination IP address to a private IP address, the routing decision is made based on the destination IP address. Just before the forwarded

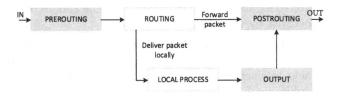

FIGURE 1.10
NAT using Linux and Netfilter.

packet leaves the machine, it passes the POSTROUTING chain and sees if the source IP address is required to be changed or not, and then leaves through the network interface. For locally generated packets, there is a small difference: Instead of passing through the PREROUTING chain, it passes the OUTPUT chain and then moves on to the POSTROUTING chain.

1.4.5 iptables

iptables is a command-line interface to the packet filtering functionality in netfilter. The packet filtering mechanism provided by iptables is organized into three different kinds of structures: tables, chains and targets. Simply put, a table is something that allows you to process packets in specific ways. The default table is the filter table, although there are other tables too. These tables have chains attached to them. These chains are used to inspect traffic at various points, such as when they just arrive on the network interface or just before they are handed over to a process. Rules can be added to them to match specific packets such as TCP packets going to port 80 and associate it with a target. A target decides the fate of a packet, such as allowing or rejecting it.

When a packet arrives (or leaves, depending on the chain), iptables matches it against rules in these chains one-by-one. When it finds a match, it jumps onto the target and performs the action associated with it. If it does not find a match with any of the rules, it simply does what the default policy of the chain tells it to. The default policy is also a target. To set up a whitelist, the default policy is DROP and to set up a blacklist, the default policy is ACCEPT. By default, all chains have a default policy of accepting packets.

1.4.5.1 Tables in iptables

There are several tables included in iptables:

- *Filter table*: This is the default and perhaps the most widely used table. It is used to make decisions about whether a packet should be allowed to reach its destination.
- *Mangle table*: This table is used to alter packet headers in various ways, such as changing Time-To-Live (TTL) values.
- *NAT table*: This table is used to route packets to different hosts on NAT networks by changing the source and destination addresses of packets. It is often used to allow access to services that cannot be accessed directly, because they are on a NAT network.
- *Raw table*: iptables is a stateful firewall, which means that packets are inspected with respect to their "state." For example, a packet could be part of a new connection, or it could be part of an existing connection. The raw table can be used to work with packets before the kernel starts tracking its state. In addition, we can also exempt certain packets from the state-tracking machinery.

In addition, some kernels also have a security table. It is used by Security-Enhanced Linux (SELinux) to implement policies based on SELinux security contexts.

1.4.5.2 Chains in iptables

Each of the tables in iptables is composed of a few default chains. These chains are used to filter packets at various points. The list of chains iptables provides is as follows:

- The PREROUTING chain: Rules in this chain apply to packets as they just arrive on the network interface. This chain is present in the NAT, mangle and raw tables.
- The INPUT chain: Rules in this chain apply to packets just before they are given to a local process. This chain is present in the mangle and filter tables.
- The OUTPUT chain: The rules here apply to packets just after they have been produced by a process. This chain is present in the raw, mangle, NAT and filter tables.
- The FORWARD chain: The rules here apply to any packets that are routed through the current host. This chain is only present in the mangle and filter tables.
- The POSTROUTING chain: The rules in this chain apply to packets as they just leave the network interface. This chain is present in the NAT and mangle tables.

The diagram in Figure 1.11 shows the flow of packets through the chains in various tables.

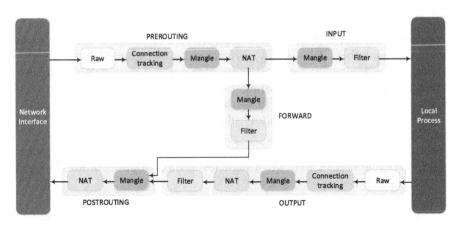

FIGURE 1.11
iptables.

1.4.5.3 Targets in iptables' Chains

Chains in iptables allow us to filter traffic by adding rules to them. So, for example, we can add a rule on the filter tables INPUT chain to match traffic on port 22. But what would we do after matching them? That is what targets are for; they decide the fate of a packet.

Some targets are terminating, which means that they decide the matched packets' fate immediately. The packet will not be matched against any other rules. The most commonly used terminating targets are:

- ACCEPT: This causes iptables to accept the packet.
- DROP: iptables drops the packet. To anyone trying to connect to your system, it would appear like the system did not even exist.
- REJECT: iptables "rejects" the packet. It sends a "connection reset" packet in case of TCP, or a "destination host unreachable" packet in case of UDP or Internet Control Message Protocol (ICMP).

Besides the above terminating targets, there are also non-terminating targets, which keep matching other rules even if a match was found. An example of this is the built-in *LOG* target. When a matching packet is received, it logs about it in the kernel logs. However, iptables keeps matching it with the rest of the rules too.

Sometimes, the system may have a complex set of rules to execute once a matched packet is found. In addition to default chains, we can also create a custom chain. In a table, it must start from a default chain, and then, it could jump to a user-defined chain.

1.5 IP Network Routing

IP network routing is the process of moving packets across a network from one host to another. It is usually performed by dedicated devices called a router, which is usually a computer running routing protocols to establish routing tables for packet forwarding. Network routing is usually referred to as IP network routing, in which the connectivity is established on IP networks. IP packets are the fundamental unit of information transport in all modern computer networks.

Routing protocols play important roles for modern networks to establish end-to-end interconnections. At a high level, most routing protocols have three components:

- *Determining neighbors*: Network nodes that are directly linked through layer-2 connections are called neighbors in a network topology. A node may not be able to reach a directly linked node either because the link has failed or because the node itself has failed for some reason. A link may fail to deliver all packets (e.g., because a blackhole cuts cables), or may exhibit a high packet loss rate that prevents all or most of its packets from being delivered. To maintain the reachability, a HELLO protocol is usually required to monitor up or down status of neighbors. The basic idea is for each node to send periodic HELLO messages on all its live links; any node receiving a HELLO knows that the sender of the message is currently alive and a valid neighbor.

- *Periodic advertisements*: Each node periodically sends routing advertisements to its neighbors. These advertisements summarize useful information about the network topology. Each node sends these advertisements periodically to achieve two functions. First, in distance vector routing protocols, periodic advertisements ensure that over time all nodes have correct information to compute correct routes. Second, in both distance-vector and link-state routing protocols, periodic advertisements are the fundamental mechanism used to overcome the effects of link and node failures. Both distance-vector and link-state routing protocols are described next.

- *Integrating advertisements*: In this step, a node processes all the advertisements it has recently heard to establish the routing table. To keep an up-to-date routing table, three continuous steps need to be maintained: discovering the current set of neighbors, disseminating advertisements to neighbors, and adjusting the routing tables. This continual operation implies that the state maintained by a router is soft, i.e., it refreshes periodically as updates arrive, and adapts to changes that are represented in these updates. This soft state means that the path between a source and destination pair could be changed at any time, which can change the order of a sequence of packets received at a destination. On the positive side, the ability to refresh the route means that the system can adapt by "rerouting" packets to deal with link and node failures.

A variety of routing protocols have been developed in the literature and several different ones are used in practice. Broadly speaking, protocols fall into one of two categories depending on what they send in the advertisements and how they integrate advertisement to compute the routing table. Protocols in the first category are called distance vector protocols, where each node n advertises to its neighbors a vector with one component per destination. In the integration step, each recipient of the advertisement can compute its vector of distance based on received vectors to update shorter distances.

Routing protocols in the second category are called link-state protocols, where each node n advertises information (in a Link-State Update (LSU) packet) about the link status (or cost) to all its current neighbors; and each recipient re-sends this information on all of its links except the interface that received the LSU, which is called flooding. Every node will repeat the flooding procedure, and eventually, all nodes will know about all the links' states as long as the network is connected. In the integration step, each node uses the shortest path algorithm, i.e., Dijkstra, to compute the minimum-cost path to every destination in the network.

Due to the flooding nature of link-state protocols, a huge amount of traffic will be generated, which make it unsuitable for large networks. In general, link-state protocols are used for intra-domain routing such as Open Shortest Path First (OSPF) [199] and Intermediate System to Intermediate System (IS-IS) [208]. On the contrary, distance vector protocols have the issue of long-time convergence; however, due to its distributed nature for routing table computation, distance vector protocols have been widely deployed for both intra-domain routing, e.g., Routing Information Protocol (RIP) [110], and inter-domain routing, e.g., Border Gateway Protocol version 4 (BGPv4) [227].

Summary

In this chapter, our discussions focus on computer network foundations. Particularly, understanding the TCP/IP protocol stack and various addresses to be used in computer networking protocols is extremely important to understand how the Internet works. These foundations allow us to further look into concepts of logical and virtual networks and how we can use various computer networking services to realize the different views of a computer network at its physical level as well as logical level.

Computer networks build the fundamental Internet infrastructure to support various Internet applications. Back to the example of sending an email vs. sending a mail given at the beginning of this chapter, we realized Internet can greatly improve the efficiency of message delivery and reduce the cost of the traditional mail service on how to deliver a letter. The major benefit of using the digitalized Internet to revitalize the traditional messaging delivery system is to remove the human-in-the-loop requirement at the intermediate nodes through the message delivery path and automate the control function for message delivery via various programmable interfaces. In the following chapters, we can see many protocol examples and they mainly focus on how to build more effective interfaces and incorporate more programmable capabilities to further improve the efficiency and functionalities of our network.

The automation at the control plane of a computer network also invokes great security concern. To avoid our computer networking system from being vandalized by malicious attackers, we will gradually introduce security services and models in the following chapters to provide a more secure and trustworthy networking environment in the fast-evolving computer networking research and development area.

2

Virtual Networking

How do things change when a physical server is converted into a Virtual Machine (VM), run on top of a Hypervisor (like KVM, VirtualBox, XEN, vSphere, Hyper-V, etc.) and connected to a virtual network? With more and more servers being virtualized (occupying the majority of all servers in the world now), we need to understand virtual network connections and virtual infrastructures.

In this chapter, we provide a review of virtual network solutions. Compared to a physical network, which is a network of physical machines that are connected; a virtual network is a technical approach that provides mapping functions to represent various logical views or isolations of underlying physical networks at the network layer of the TCP/IP protocol stack to better serve its supported applications for easier management and stronger security protection.

In this chapter, we first provide a few definitions to clarify the differences among physical networks, virtual networks, overlay networks, and logical networks in Section 2.1. Next, we focus on layer-2 virtual networking techniques and realizations in Section 2.2. Then, in Section 2.3, we describe a closely related concept – Virtual Private Network (VPN)-at different layers of the protocol stack. Finally, we briefly discuss virtual networking and forwarding concepts in Section 2.4.

2.1 Virtual Networks

2.1.1 Basis of Virtual Networks

A virtual network is a mapping of the entire or subset of networking resources to a specific protocol layer. Virtual networking technologies enable creation of various logical networks based on the applications' need. For example, a virtual network interconnects virtual and/or physical machines that are connected to each other so that they can send data to and receive data from each other. Generally speaking, two virtual machines residing on the same physical server can send data to each other through a virtual network, e.g., a software switch running on top of the operating system. As shown in Figure 2.1(a), when two virtual machines located on different physical servers and want to send data to each other through a software switch, a layer-2

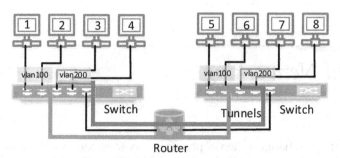

(*a*) Physical Network with Layer-2 Tunnels (e.g., GRE/VxLAN).

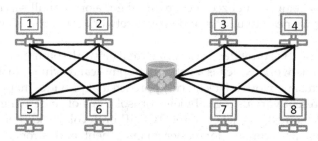

(*b*) Logical Network (Link-layer View).

FIGURE 2.1
Virtual networking approaches enable different logical views of the underlying physical networks.

tunnel needs to be established overlaying the physical network between two servers. Thus, it is important to understand the concept of layered structure of TCP/IP protocols; and how we can create a virtual network by using virtual networking parameters at different protocol layers and how we can create a virtual network using overlaying approaches. The virtual networks changed logical networks as shown in Figure 2.1(b) vs. the logical networks shown in Figure 1.7(b).

Virtual networks can be achieved through Network Virtualization (NV) approach, which is defined by the ability to create logical, virtual networks that are decoupled from the underlying network hardware to ensure the network can better integrate with and support increasingly virtual environments. NV can be delivered via hardware into a logical virtual network that is decoupled from and runs independently on top of a physical network. Beyond Layer 2 or 3 (L2 or L3) services like switching and routing, NV can also incorporate virtualized L4-7 services.

With virtualization, organizations can take advantage of the efficiencies and agility of software-based compute and storage resources. While networks have been moving towards greater virtualization, it is only recently, with the true decoupling of the control and forwarding planes, as advocated by

SDN [153] and Network Functions Virtualization (NFV) [106], that network virtualization has become more of a focus.

The two most common forms of NVs are protocol-based virtual networks, (1) Virtual LANs (VLANs) [1], Virtual Networks (VNs) and Virtual Private Networks (VPNs) [25], Virtual Private LAN Services (VPLS) [168], and (2) virtual networks that are based on virtual devices (such as the networks connecting VMs inside a Hypervisor). In practice, both forms can be used in conjunction.

VLANs are logical local area networks (LANs) based on physical LANs as shown in Figure 2.1(*a*). A VLAN can be created by partitioning a physical LAN into multiple logical LANs using a VLAN ID. Alternatively, several physical LANs can function as a single logical LAN. The partitioned network can be on a single router, or multiple VLANs can be on multiple routers just as multiple physical LANs would be.

A VPN is usually built on tunneling protocols, which consists of multiple remote end-points (typically routers, VPN gateways of software clients) joined by some sort of tunnel over another network, usually a third-party network. Two such end-points constitute a "Point to Point Virtual Private Network" (or a PTP VPN). Connecting more than two end points by putting in place a mesh of tunnels creates a 'Multipoint VPN'.

A VPLS is a specific type of Multipoint VPN. VPLS are divided into Transparent LAN Services (TLS) and Ethernet Virtual Connection Services. A TLS sends what it receives, so it provides geographic separation, but not VLAN subnetting. An Ethernet Virtual Connections (EVCS) adds a VLAN ID, so it provides geographic separation and VLAN subnetting.

A common example of a virtual network that is based on virtual devices is the network inside a Hypervisor where traffic between virtual servers is routed using virtual switches (vSwitches) along with virtual routers and virtual firewalls for network segmentation and data isolation. Such networks can use non-virtual protocols such as Ethernet as well as virtualization protocols such as the VLAN protocol IEEE 802.1Q [129].

2.1.2 Abstraction vs. Virtualization

Virtual networks are built on virtualization techniques and virtualization is an important technique for establishing modern cloud computing services. However, it is easy to confuse with another overly used concept – *abstraction*. Virtualization is similar to abstraction but it does not always hide the low layer's details. A real system is transformed so that it appears to be different. Moreover, virtualization can be applied not only to subsystem, but to an entire machine, e.g., Virtual Machines (VMs), or an entire system, e.g., virtual networks. Abstraction is about hiding details, and involves constructing interfaces to simplify the use of the underlying resource (e.g., by removing details of the resource's structure). For example, a file on a hard disk that is mapped to a collection of sectors and tracks on the disk. We usually do not directly address disk layout when accessing the file. *Concrete* is the opposite of abstract. For example, software and development goes from concrete, e.g., the actual binary instructions, to abstract, e.g., assembly to *C* to *Java* to a framework like *Apache Groovy* [31] to a customizable *Groovy add-on*. For computer virtualization solutions, hardware no longer exists for the OS. At some level it does, on the host system, but the host system creates a virtualization layer that maps hardware functions that allows an OS to run on the software rather than the hardware. For network virtualization, it focuses on creating an overplayed networking solution to logically isolate multiple networks that are physically sharing the same set or subset of networking resources. For networking resource management, how to allocate virtual networks on physical networks is called virtual networking embedding, which will be discussed in the next subsection.

Besides abstraction, the term *overlay network* is also quite often mentioned, in which it describes a computer network that is built on top of another network. Nodes in the overlay network can be thought of as being connected by logical links, each of which corresponds to a path, perhaps through many physical links, in the underlying network. For example, distributed systems such as peer-to-peer networks are overlay networks because their nodes run on top of the Internet. The Internet was originally built as an overlay upon the telephone network, while today (through the advent of Voice over IP (VoIP)), the telephone network is increasingly turning into an overlay network built on top of the Internet. Overlay networks build the foundation of virtual networks and they run as independent virtual networks on top of a physical network infrastructure. Virtual network overlays allow resource providers, such as cloud providers, to provision and orchestrate networks alongside other virtual resources. They also offer a new path to converged networks and programmability.

Two additional concepts are also quite frequently used with the concept of virtualization: *replication* is to create multiple instances of the resource (e.g., to simplify management or allocation); and *isolation* is to separate the uses which clients make of the underlying resources (e.g., to improve security).

2.1.3 Benefits of Virtualizing Networks

In computing, **virtualization** refers to the act of creating a virtual (rather than actual) version of something, including virtual computer hardware platforms, storage devices, and computer network resources. Generally speaking, virtualization is a technical approach to create illusions, which is dominated by three major virtualized resources, computing power, networking resources, and storage spaces. A virtual network is a computer network that consists, at least in part, of virtual network links. Usually, a virtual network link is a link that does not consist of a physical (wired or wireless) connection between two computing devices but is implemented using methods of NFV. Thus, a virtualized system's interface and resources are mapped onto interface and resources of another ("real") system; and virtualization provides a different interface and/or resources at a similar level of the abstraction without losing major functions to be represented in the virtualized system. Focusing on network virtualization, several benefits are summarized as follows:

- *Resource optimization*: For satisfying real needs, the available network bandwidth can be used much more effectively. Tunnels and virtual networks can isolate or restrict network traffic based on the application needs, which provides in-network control capability compared to traditional best effort computer networking solutions.

- *Multiple execution environments*: Virtual networks enable an isolated multiple application running environment that can bring multifaceted benefits such as isolating or restricting malicious network traffic, providing redundancy for backup, load sharing features, etc.

- *Debugging and intrusion detection*: Using virtual networks, we can create slices (called slicing), which is an isolation approach to reserve a certain amount of networking resources for a particular application. In case of under-attacks, a slice can be handled easily such as performing intrusion detection, monitoring, and Deep Packet Inspection (DPI), etc.

- *Mobility*: Virtual networking not only provides a programmable approach to setup a needed network for new service deployment and management, but also it can seamlessly assist software migration. For example, live migration consists of migrating the memory content of the VM, maintaining access to the existing storage infrastructure containing the VM's stored content while providing continuous network connectivity to the existing network domain. The existing VM transactions can be maintained and any new transaction will be allowed during any stage of the live migration, thereby providing continuous data availability.

- *Appliance (security)*: Enabling security function plug-and-play feature and working as appliances that can be easily enabled and disabled, and applied at any network segment or interfaced to existing services in a cloud-based service platform.

- *Testing/Quality assurance*: Virtual networks offer software developers isolated, constrained, test environments. Rather than purchasing and maintaining dedicated physical hardware, virtual networks can create multiple isolated networking environments. Together with VMs and virtual storages, we can quickly establish a software testing environment, where developers can create an unlimited number of user configurations on their physical machines and choose the most suitable configuration at each stage. This gives the possibility to experiment with potentially incompatible applications and perform testing with different user profiles. Moreover, the testing will be hardware independent, portable, and backup and replication will be easy.

2.1.4 Orchestration and Management of Virtual Networks

We are in the age of virtualization, associating network resources and elements to services is no longer as simple as literally running wires and connecting terminal blocks. Nor does it just mean setting new parameters so network hardware knows how to provide the customer's VPN, for example. A virtual assembly process needs to be included into operations practices, which describe how abstract services, features and even devices are realized on real infrastructure.

At the same time, as IT and the network merge in the cloud, traditional operations support systems, business support systems and network management systems may be merging into a cloud-related operations model. That model focuses on the emerging concepts of management and orchestration, and of managing bindings, which are records of the physical and virtual elements that make up a service.

SDN and Network Functions Virtualization (NFV) are new computer networking management frameworks, described in Chapter 3.

2.1.5 Virtual Networking Embedding Problems

In network virtualization, the primary entity is a virtualized network, which is a combination of active and passive network elements (network nodes and network links) on top of a substrate network. Virtual nodes are interconnected through virtual links, forming a virtual topology. By virtualizing both node and link resources of a substrate network, multiple virtual network topologies with widely varying characteristics can be created and co-hosted on the same physical hardware. Moreover, the abstraction introduced by the resource virtualization mechanisms allows network operators to manage and modify networks in a highly flexible and dynamic way.

How to optimally allocate virtual networks and their associated networking resources is called a Virtual Network Embedding (VNE) problem, which is an NP hard problem. Through dynamic mapping of virtual resources onto physical hardware, the benefit gained from existing hardware can be maximized. Optimal dynamic resource allocation, leading to the self-configuration and organization of future networks, will be necessary to provide customized end-to-end guaranteed services to end users. This optimality can be computed with regard to different objectives, ranging from QoS, economical profit, or survivability over energy efficiency to security of the networks.

VNE deals with the allocation of virtual resources both in nodes and links. Therefore, it can be divided in two sub problems: Virtual Node Mapping (VNoM) where virtual nodes have to be allocated in physical nodes and Virtual Link Mapping (VLiM) where virtual links connecting these virtual nodes have to be mapped to paths connecting the corresponding nodes in the substrate network.

2.1.5.1 VNE Problem Description

The application of virtualization mechanisms to network resources leads to the question how the virtualized resources should be realized by the substrate resources. It is important to note that substrate resources can be virtual themselves. This is commonly referred to as nested virtualization. In that case, only the lowest layer has to consist of physical resources.

The hardware abstraction provided by the virtualization solution provides a common denominator, allowing any substrate resource to host virtual resources of the same type. Typically, a substrate resource is partitioned to host several virtual resources. For example, a virtual node can, in principle, be hosted by any available substrate node. Moreover, a single substrate node can host several virtual nodes. Thus, the mapping of virtual nodes to substrate nodes describes a $n : 1$ relationship (a strict partition of substrate resources).

In some cases, substrate resources can also be combined to create new virtual resources. This is the case for a virtual link which spans several links (i.e., a path) in the substrate network. In this case, a virtual link between two virtual nodes v and w is mapped to a path in the substrate network that connects the substrate hosts of v and w. Each substrate link may then be part of several virtual links. As such, the mapping of virtual links to substrate paths describes a $n : m$ relationship (both, a partition and a combination of substrate resources).

Figure 2.2 depicts a scenario where three virtual networks with 2-3 nodes each are hosted on one substrate network with four nodes. It can be seen that substrate nodes can host several virtual nodes (up to three in this

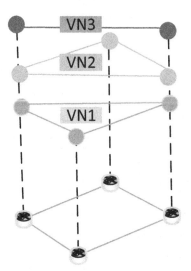

FIGURE 2.2
A VNE example.

example). Likewise, substrate links can host more than one virtual link. Moreover, three of the virtual links span two substrate links, thus representing a virtual resource combined from several substrate resources.

In general, there are some restrictions to be considered during the mapping. The candidate substrate resources for a mapping have to be able to support the performance requirements of the virtual resources. For example, a 1000 MBit/s virtual link cannot be mapped to a path containing a 100 MBit/s substrate link. Likewise, the Central Processing Unit (CPU) power requested by a virtual node has to be less than (or equal to) the CPU power actually provided by a substrate node. If redundancy is required, even more substrate resources may have to be reserved. Nevertheless, substrate resources should be spent economically. Therefore, the mapping has to be optimized. This problem of mapping virtual resources to substrate resources in an optimal way is commonly known as the VNE problem. This is typically modeled by annotating a Virtual Network Request (VNR) with node and link demands. Likewise, the Substrate Network (SN) is annotated with node and link resources. Demands and resources then have to be matched in order to complete the embedding. This means that virtual resources are first mapped to candidate substrate resources. Only if all virtual resources can be mapped, the entire network is then embedded and substrate resources are actually spent. If VNRs arrive one at a time, reconfiguration might be necessary, reverting the previous embedding and calculating a new mapping.

2.1.5.2 VNE Formal Definition

Here, we use the formal definition of VNE presented in [91]. The VNE problem can be described as follows (see Table 2.1): Let $SN = (N, L)$ be a substrate network where N represents the set of substrate nodes and L the set of

TABLE 2.1

Notations for VNE

Term	Description
$SN = (N, L)$	SN is a substrate network, consisting of nodes N and links L
$VNR^i = (N^i, L^i)$	VNR^i denotes the ith Virtual Network Request (VNR), consisting of nodes N^i and links L^i
$\dot{R} = \prod_{j=1}^{m} R_j$	$\dot{R}$ contains resource vectors for all resources $R_1, \ldots, R_m$
$cap : N \cup L \to \dot{R}$	The function cap assigns a capacity to an element of the substrate network (either node or link)
$dem_i : N^i \cup L^i \to \dot{R}$	The function demi assigns a demand to an element of VNR^i (either a node or a link)
$f_i : N^i \to N$	f_i is the function that maps a virtual node of VNR^i to a substrate node (VNoM)
$g_i : L^i \to SN' \subseteq SN$	g_i is the function that maps a virtual link of VNR^i to a path in the substrate network (VLiM)

substrate links and let $VNR^i = (N^i, L^i)$ be a set of $i = 1, \ldots, n$ VNRs where N^i and L^i represent the set of virtual nodes and virtual links of the VNR i, respectively. Furthermore, let $\dot{R} = \sum_j^m = 1 R_j$ be a vector space of resource vectors over resource sets $R_1, \ldots, R_m$ and let cap : $N \cup L \rightarrow \dot{R}$ be a function that assigns available resources to elements of the substrate network. Finally, for each VNR^i, let $dem_i : N^i \cup L^i \rightarrow \dot{R}$ be a function that assigns demands to elements of all VNRs. Then, a VNE consists of two functions $f_i : N^i \rightarrow N$ and $g_i : L^i \rightarrow SN' \subseteq SN$ for each VNR^i such that $\forall n^i \in N^i : dem_i(n^i) \leq cap(f_i(n^i))$ and $\forall l^i \in L^i : \forall l \in g_i(l^i) : dem_i(l^i) \leq cap(l)$. f_i is then called a node mapping function (VNoM) and g_i is called a link mapping function (VLiM). Together, they form an embedding for VNR^i. It is not required that these functions are calculated by a single entity; calculation can be split among multiple entities.

Solving the VNE problem is NP-hard, as it is related to the multi-way separator problem [30]. Even with a given virtual node mapping, the problem of optimally allocating a set of virtual links to single substrate paths reduces to the unsplittable flow problem [155, 158], and thus also is NP hard. Therefore, truly optimal solutions can only be gained for small problem instances. Thus, currently the main focus of work within the research community is on heuristic or metaheuristic approaches.

VNE is a central problem to be solved when networks are virtualized. Optimizing the embedding of multiple virtual networks on one substrate network is computationally difficult for a number of important metrics. Multiple algorithms approaching this problem have been discussed in the literature, which can be reviewed in a survey paper [91].

2.2 Layer-2 Virtual Networking

Based on OSI model, layer-2 networking provides direct connectivity between hosts, where the TCP/IP networking model the direct connectivity is provided at the Interface layer (or represented as a sublayer of the Physical layer). Before presenting the layer-2 virtual networking solutions, it is important to understand the difference among a few networking devices and their functions at different protocol layers.

A **hub** is the simplest of networking devices. We rarely see hubs in our current networking environment due to its inefficiency of data delivery. However, due to its simplicity and critical role served in the history of computer networking, we should understand what the difference between a hub and a switch is. In the OSI networking model, a hub is a physical-layer device, and it is the central part of a wheel where the spokes come together. In a hub, a frame is passed along or *broadcast* to every one of its ports. It does not matter that the frame is only destined for one port. The hub has no way of distinguishing which port a frame should be sent to, and thus, it passes a frame to every port to ensure that it will reach its intended destination. Moreover, hubs cannot filter data so data packets are sent to all connected devices/ computers and do not have intelligence to find out the best path for data packets. This leads to inefficiencies and wastage. As the result, hubs are used on small networks where data transmission is not high.

Compared to hub, a **switch** is a data-link layer device, which has more control capabilities on data frames. A switch keeps a record of the MAC addresses of all the devices connected to it. With this information, a switch can identify which system is sitting on which port. As the result, when a frame is received, the switch knows exactly which port to send it to, without significantly increasing network response times. Unlike a hub, all the connected devices share the bandwidth of the full capacity of the hub; the switch can grant the full capacity of the switch to each individual connected device due to the direct connections established between switch ports. For example, for an 8-port hub with the bandwidth capacity of 100Mbps, each device can theoretically get the 1/8 portion of the bandwidth; however, for a switch, each port can achieve the full 100Mbps transmission rate. Moreover, switches may have more control functions such as performing error checking before forwarding data and creating virtual LANs, and some advanced switches can also handle IP packets (i.e., at the network layer/Layer-3).

In the physical world, a bridge connects roads on separate sides of a river or railroad tracks. In the technical world, a **bridge** is also a layer-2 (i.e., data-link layer) device. A bridge connects a LAN to another LAN that uses the same protocol, i.e., a bridge is a device that separates two or more network segments within one logical network (e.g., a single IP-subnet). A bridge examines each message on a LAN, *passing* those known to be within the same LAN and forwarding those known to be on other interconnected LANs. In bridging networks, computer or node addresses have no specific relationship to location. For this reason, frames are sent out to every address on the network and accepted only by the intended destination node. Bridges learn which addresses are on which network and develop a learning table so that subsequent messages can be forwarded to the right network. Bridges use Spanning Tree Protocol (STP) developed to send frames using broadcasting and multicasting without causing frame storm issues

The challenge here is *how to differentiate between a switch and a bridge*? Before describing the difference, we need to first explain how Ethernet works, in which switches and bridges are designed to support. Ethernet was originally an *everyone sees all traffic* protocol running over a LAN. When a node is using the network, others need to wait until the next chance that no one uses the network. If two nodes try to use the network at the same time, then collision occurs, and then both nodes need to wait a random amount of time before attempting to use the network again. This illustrates that Ethernet is designed for a shared communication media (or bus), and each Ethernet is a *collision domain* or what is called *broadcast domain*. Traditionally, a bridge usually is explained as a 2-port device to connect two LANs, where we consider each LAN as a shared bus or collision domain. Using a bridge, a large collision domain can be separated into two smaller collision domains, and thus reducing the chance of collisions for each port. From this view, we can simply illustrate that an Ethernet switch is a multi-port Ethernet bridge, i.e., a switch is simply a bridge with lots of ports. Thus, we can simply view a switch is the same device as a multi-port bridge. A switch will increase the number of collision domains on a LAN, where each port will be a collision domain; however, the size of these domains will be reduced. For a full-duplex switch, technically, the switch can remove collisions totally.

A **router** is a network-layer device (layer-3 device according to the OSI model). Unlike a switch forwarding frames based on MAC addresses, a router forwards packets based on IP addresses and routing tables established using dynamic routing protocols or static routes. Moreover, unlike a switch interconnecting networks or nodes at the data-link layer, a router interconnects different IP networks.

2.2.1 Linux Bridge

As described in previous section, a switch can be considered as a multi-port bridge. A switch is responsible for connecting several network links to each other, creating a Local Area Network (LAN). In this book, without a special notice, we do not differentiate between the term *switch* and *bridge*. Generally speaking, a switch is composed by four major components: a set of network ports, a control plane, a forwarding plane, and a MAC learning database.

The set of ports are used to interconnect interfaced networking devices and hosts and forward traffic between other switches and end-hosts in the LAN. The control plane of a switch is typically used to run the Spanning Tree Protocol (STP) [13] and calculate a Minimum Spanning Tree (MST) [100] for the LAN, preventing physical loops from crashing the network. The forwarding plane is responsible for processing input frames from the network ports and making a forwarding decision on which network ports to forward a received frame. The MAC learning database is used to keep track of the host locations in the LAN. It typically contains an entry for each host MAC address that traverses the switch, and the input port where the frame was received.

For each unicast destination MAC address, the switch looks up the output port in the MAC database. If an entry is found, the frame is forwarded through the port further into the network. If an entry is not found, the frame is instead flooded from all other network ports in the switch, except the port where the frame was received.

Several operating systems have their own bridging implementation in their network stack. For example, FreeBSD [186] has a similar bridging implementation to Linux kernel; however, the FreeBSD implementation also implements the Rapid Spanning Tree Protocol (RSTP) [280]. The FreeBSD bridge implementation also supports more advanced features, such as port MAC address limits, and Simple Network Management Protocol (SNMP) monitoring of the bridge state.

In the following subsections, we present the architecture, design and the implementation of the Linux bridging module. The architectural overview of the Linux bridging module is divided into three parts: (1) data structures for the bridging module, (2) the configuration interface of the Linux bridging module, and (3) the input/output processing flow of the Linux bridging module.

2.2.1.1 Data Structures of Linux Bridge

The Linux bridge module has three key data structures that provide the central functionality for the bridge operation [271]. Figure 2.3 presents an overview of the most important fields and their associations in the three key data structures. The main data structure for each bridge in the operating system is the *net_bridge*. It holds all of the bridge-wide configuration information, a doubly linked list of bridge ports (*net_bridge_port* objects) in the field *port_list*, a pointer to the bridge *netdevice* in the field *dev*, and the forwarding database in the field *hash*. Finally, the field *lock* is used by the bridge to synchronize configuration changes, such as port additions, removals, or changing the various bridge-specific parameters.

Each bridge port has a separate data structure *net_bridge_port*. It contains the bridge port specific parameters. The field *br* has a back reference to the bridge, which the port belongs to. The *dev* field holds the actual network interface that the bridge port uses to receive and transmit frames. Finally, position of the data structure object in the *net_bridge → port_list* linked list is stored in the field *list*.

The third key data structure for the Linux bridge module is the *net_bridge_fdb_entry* object that represents a single forwarding table entry. A forwarding table entry consists of a MAC address of the host (in the field *addr*), and the port where the MAC address was last seen (in the field *dst*). The data structure also contains a field *hlist* that points back to the position of the object in a hash

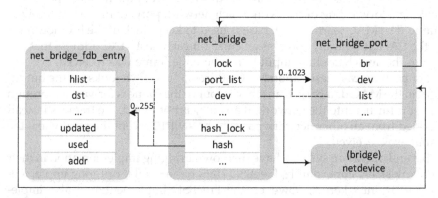

FIGURE 2.3
Linux bridge data structure.

table array element in *net_bridge→hash*. In addition, there are two fields, *updated* and *used*, which are used for timekeeping. The *updated* field specifies the last time when the host was seen by this bridge, and the *used* field specifies the last time when the object was used in a forwarding decision. The *updated* field is used to delete entries from the forwarding database, when the maximum inactivity timeout value for the bridge is reached, i.e., *current_time-updated* > *bridge_hold_time*.

2.2.1.2 Linux Bridge Configuration

The Linux bridging module has two separate configuration interfaces exposed to the user-space. The first, *ioctl* interface offers an interface that can be used to create and destroy bridges in the operating system, and it can also add network interfaces and remove existing network interfaces to/from the bridge. The second is the *sysfs*-based interface that allows the management of bridge and bridge port specific parameters. Figure 2.4 presents a high level overview of the kernel *ioctl* process, which creates and initializes the bridge object, and adds network interfaces to it.

The creation of a new bridge begins with the *ioctl* command *SIOCBRADDBR* that takes the bridge interface name as a parameter. The *ioctl* command is handled by the *br_ioctl_deviceless_stub* function, as there is no bridge device to attach the *ioctl* handler internally. The addition of a new bridge calls the function *br_add_bridge*, that creates required bridge objects in the kernel, and eventually calls the *alloc_netdev* function to create a new *netdevice* for the bridge. The allocated *netdevice* is then initialized by the *br_dev_setup* call, including assigning the bridge device specific *ioctl* handler *br_dev_ioctl* to the newly allocated *netdevice*. All subsequent bridge specific *ioctl* calls are done on the newly created bridge device object in the kernel.

Ports are added to bridges by the *ioctl* command *SIOCBRADDIF*. The *ioctl* command takes the bridge device and the index of the interface to add to the bridge as parameters. The *ioctl* calls the bridge device *ioctl* handler

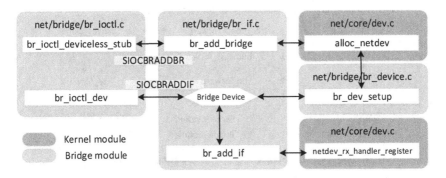

FIGURE 2.4
Linux bridge configuration: adding a bridge and a bridge port.

(*br_dev_ioctl*), that in turn calls the *br_add_if* function. The function is responsible for creating and setting up a new bridge port by allocating a new *net_bridge_port* object. The object initialization process automatically sets the interface to receive all traffic, adds the network interface address for the bridge port to the forwarding database as a local entry, and attaches the interface as a slave to the bridge device. Finally, the function calls the *netdev_rx_handler_register* function that sets the *rx_handler* of the network interface to *br_handle_frame*, that enables the interface to start processing incoming frames as a part of the bridge.

2.2.1.3 Linux Bridge Frame Processing

The Linux bridge processing flow begins from lower layers. Each network interface that acts as a bridge interface has a *rx_handler* set to *br_handle_frame*, which acts as the entry point to the bridge frame processing code. The *rx_handler* is called by the device-independent network interface code in *__netif_receive_skb*. Figure 2.5 presents the processing flow of an incoming frame, as it passes through the Linux bridge module to a destination network interface queue.

The *br_handle_frame* function does the initial processing on the incoming frame. This includes doing initial validity checks on the frame and separating control frames from normal traffic. The bridge considers any frame that has a destination address prefix of 01:80:C2:00:00 to be a control frame, that may need specialized processing. The last byte of the destination MAC address defines the behavior of the link local processing. Ethernet pause frames are automatically dropped. STP frames are either passed to the upper layers if it is

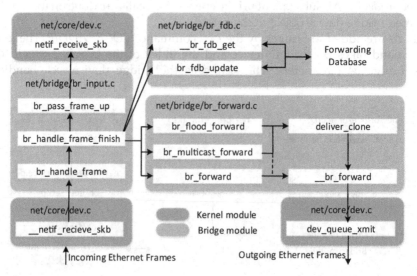

FIGURE 2.5
Linux bridge modules and I/O.

enabled on the bridge or forwarded when it is disabled. If a forwarding decision is made, and the bridge is in either forwarding or learning mode, the frame is passed to *br_handle_frame_finish*, where the actual forwarding processing begins.

The *br_handle_frame_finish* function first updates the forwarding database of the bridge with the source MAC address, and the source interface of the frame by calling *br_fdb_update* function. The update either inserts a new entry into the forwarding database or updates an existing entry.

The processing behavior is decided based on the destination MAC address in the Ethernet frame. Unicast frames will have the forwarding database indexed with the destination address by using the *br_fdb_get* function to find out the destination *net_bridge_port* where the frame will be forwarded to. If a *net_bridge_fdb_entry* object is found, the frame will be directly forwarded through the destination interface by the *br_forward* function. If no entry is found for the unicast destination Ethernet address, or the destination address is broadcast, the processing will call the *br_flood_forward* function. Finally, if the frame is a multi-destination frame, the multi-cast forwarding database is indexed with the complete frame. If selective multi-casting is used and a multi-cast forwarding entry is found from the database, the frame is forwarded to the set of bridge ports for that multi-cast address group by calling the *br_multicast_forward* function. If no entry is found or selective multi-casting is disabled, the frame will be handled as a broadcast Ethernet frame and forwarded by the *br_flood_forward* function.

In cases where the destination MAC address of the incoming frame is multi-cast or broadcast, the bridge device is set to receive all traffic, or the address is matches one of the local interfaces, a clone of the frame is also delivered upwards in the local network stack by calling the *br_pass_frame_up* function. The function updates the bridge device statistics and passes the incoming frame up the network stack by calling the device independent *netif_receive_skb* function, ending the bridge specific processing for the frame.

The forwarding logic of the Linux bridge module is implemented in three functions: *br_forward*, *br_multicast_forward*, and *br_flood_forward* to forward unicast, multi-cast, and broadcast or unknown unicast destination Ethernet frames, respectively. The *br_forward* function checks whether the destination bridge interface is in forwarding state, and then either forwards the incoming frame as is, clones the frame and forwards the cloned copy instead by calling the *deliver_clone* function, or doing nothing if the bridge interface is blocked. The *br_multicast_forward* function performs selective forwarding of the incoming Ethernet frame out of all of the bridge interfaces that have registered multi-cast members for the destination multi-cast address in the Ethernet frame, or on interfaces that have multi-cast routers behind them. The *br_flood_forward* function iterates over all of the interfaces in the bridge and delivers a clone of the frame through all of them except the originating interface. Finally, all three types of forwarding functions end up calling the *br_forward* function that actually transfers the frame to the lower layers by calling the *dev_queue_xmit* function of the interface.

2.2.1.4 Use Cases of Linux Bridge

Creating a bridge

Bridge can be created using *ioctl* command *SIOCBRADDBR*, which is implemented by *brctl* utility provided by *bridge-utils*.

```
#> sudo strace brctl addbr br1
execve("/sbin/brctl", ["brctl", "addbr", "br1"], [/* 16 vars */]) = 0
...
ioctl(3, SIOCBRADDBR, 0x7fff2eae9966) = 0
...
```

There is no device at this point to handle the *ioctl* command. The *ioctl* command is handled by a stub method: *br_ioctl_deviceless_stub*, which in turn calls *br_add_bridge*. This method calls *alloc_netdev*, which is a macro that eventually calls *alloc_netdev_mqs*.

```
br_ioctl_deviceless_stub
    | - br_add_bridge
      | - alloc_netdev
        | - alloc_netdev_mqs // creates the network device
          | - br_dev_setup // sets br_dev_ioctl handler
```

alloc_netdev initializes the new *netdevice* using the *br_dev_setup*. This also includes setting up the bridge specific *ioctl* handler. In codes of the *handler*, it handles *ioctl* command to add/delete interfaces.

```
int br_dev_ioctl(struct net_device *dev, struct ifreq *rq, int cmd) {
    ...
    switch(cmd) {
      case SIOCBRADDIF:
      case SIOCBRDELIF:
      return add_del_if(br, rq→ifr_ifindex, cmd == SIOCBRADDIF);
      ...
    }
    ..
}
```

Adding an interface

In the function *br_dev_ioctl*, bridge can be created using *ioctl* command *SIOCBRADDIF*:

```
#>: sudo strace brctl addif br0 veth0
execve("/sbin/brctl", ["brctl", "addif", "br0", "veth0"], [/* 16 vars */]) = 0
...
# gets the index number of virtual ethernet device.
ioctl(4, SIOCGIFINDEX, ifr_name="veth0", ifr_index=5) = 0
close(4)
# add the interface to bridge.
ioctl(3, SIOCBRADDIF, 0x7fff75bfe5f0) = 0
...
```

br_add_if

The *br_add_if* method creates and sets up the new interface/port for the bridge by allocating a new *net_bridge_port* object. The object initialization is particularly interesting, as it sets the interface to receive all traffic, adds the network interface address for the new interface to the forwarding database as the local entry, and attaches the interface as the slave to the bridge device.

```
/* Truncated version */
int br_add_if(struct net_bridge *br, struct net_device *dev)
{
    struct ne_bridge_port *p;
    /* Don't allow bridging non-Ethernet like devices */
    ...
    /* No bridging of bridges */
    ...
    p = new_nbp(br, dev);
    ...
    call_netdevice_notifiers(NETDEV_JOIN, dev);
    err = dev_set_promiscuity(dev, 1);
    err = kobject_init_and_add(&p→kobj, &brport_ktype, &(dev→dev.kobj),
        SYSFS_BRIDGE_PORT_ATTR);
    ...
    err = netdev_rx_handler_register(dev, br_handle_frame, p);
    /* Make entry in forwarding database*/
    if (br_fdb_insert(br, p, dev→dev_addr, 0))
        ...
    ...
}
```

In the bridge implementation, since a bridge is a layer-2 device, thus, only Ethernet like devices can be added to bridge. This implementation does not add a bridge to an existing bridge. In order to set the bridge to promiscuous mode, we need to set: *de_set_promiscuity(dev, 1)*. The promiscuous mode can be confirmed from kernel logs.

```
#> grep -r 'promiscuous' /var/log/kern.log
kernel: [5185.751666] device veth0 entered promiscuous mode
```

Finally, *br_add_if* method calls *netdev_rx_handler_register*, that sets the *rx_handler* of the interface to *br_handle_frame*, and then an interface (or port) is set up in the bridge.

Linux Bridge Frame Processing

Linux bridge frame processing starts with device-independent network code in *__netif_receive_skb*, which calls the *rx_handler* of the interface and it was set to *br_handle_frame* at the time of adding the interface to bridge.

The *br_ handle_frame* does the initial processing and any address with prefix 01-80-C2-00-00 is a control plane address, that may need special processing. From the comments in *br_handle_frame*:

```
/*
 * See IEEE 802.1D Table 7-10 Reserved addresses
 *
 *      Value                      Assignment
 * 01-80-C2-00-00-00              Bridge Group Address
 * 01-80-C2-00-00-01              (MAC Control) 802.3
 * 01-80-C2-00-00-02              (Link Aggregation) 802.3
 * 01-80-C2-00-00-03              802.1X PAE address
 *
 * 01-80-C2-00-00-0E              802.1AB LLDP
 *
 * Others are reserved for future standardization
 */
```

In the method, STP messages are either passed to upper layers or forwarded if STP is enabled on the bridge or disabled, respectively. If a forwarding decision is made, the packet is passed to *br_handle_frame_finish*, where the actual forwarding happens. The highly truncated version of *br_handle_frame_finish* is provided as follows:

```
/* note: already called with rcu_read_lock */
int br_handle_frame_finish(struct sk_buff *skb)
{
    struct net_bridge_port *p = br_port_get_rcu(skbrightarrowdev);
    ...
    /* insert into forwarding database after filtering to
       avoid spoofing */
    br = prightarrowbr;
    br_fdb_update(br, p, eth_hdr(skb)rightarrowh_source, vid);
    if (prightarrowstate == BR_STATE_LEARNING)
        goto drop;
    /* The packet skb2 goes to the local host (NULL to skip). */
    skb2 = NULL;
    if (brrightarrowdevrightarrowflags & IFF_PROMISC)
        skb2 = skb;
    dst = NULL;
    if (is_broadcast_ether_addr(dest))
        skb2 = skb;
    else if (is_multicast_ether_addr(dest)) {
        ...
    } else if ((dst = __br_fdb_get(br, dest, vid)) &&
            dstrightarrowis_local) {
        skb2 = skb;
        /* Do not forward the packet since it's local. */
        skb = NULL;
    }
    if (skb) {
        if (dst) {
```

```
        br_forward(dstrightarrowdst, skb, skb2);
    } else
        br_flood_forward(br, skb, skb2);
    }
    if (skb2)
        return br_pass_frame_up(skb2);
out:
    return 0;
    ...
}
```

In the *br_handle_frame_finish*, an entry in forwarding database is updated for the source of the frame. If the destination address is a multi-cast address, and if the multi-cast is disabled, the packet is dropped, or else message is received using *br_multicast_rcv*; if the promiscuous mode is on, packet will be delivered locally, irrespective of the destination.

For a unicast address, the port is determined by using the forwarding database (*__br_fdb_get*). If the destination is local, then *skb* is set to *null*, i.e., packet will not be forwarded; if the destination is not local, then based on if we found an entry in forwarding database, either the frame is forwarded (*br_forward*) or flooded to all ports (*br_flood_forward*).

A packet is delivered locally (*br_pass_frame_up*) if either the current host is the destination or the net device is in promiscuous mode. The *br_forward* method either clones and then delivers (if it is also to be delivered locally, by calling *deliver_clone*), or directly forwards the message to the intended destination interface by calling *__br_forward*. The method *bt_flood_forward* forwards the frame on each interface by iterating through the list in the *br_flood* method.

A bridge can be used to create various different network topologies and it forms the foundation to realize more advanced networking solutions such as virtual switches, and bridges are used with containers where they provide networking in network *namespaces* along with *veth* devices. In fact, the default networking in Docker is provided using bridge.

2.2.2 Open Virtual Switches

Open vSwitch (OVS) is a multi-layer software switch licensed under the open source Apache 2 license. OVS is well suited to function as a virtual switch in VM environments. In addition to exposing standard control and visibility interfaces to the virtual networking layer, it was designed to support distribution across multiple physical servers. OVS supports multiple Linux-based virtualization technologies including Xen/XenServer, KVM, and VirtualBox.

2.2.2.1 Linux Bridge vs. Open Virtual Switch

As we described in the previous section, Linux bridge is a native function on Linux kernel with layer-2 capabilities, which can be considered as an

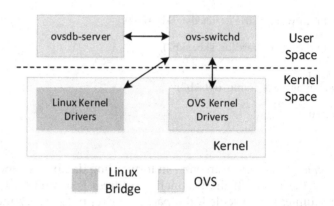

FIGURE 2.6
Open virtual switch and Linux bridge.

Ethernet Hub. OVS provides a "software layer" in the user space to add programmable capabilities involving database, layer-2 to layer-4 matching, QoS, etc., which is shown in Figure 2.6. Moreover, OVS supports OpenFlow standards [259].

2.2.2.2 Open Virtual Switch Supporting Features

OVS version 2.9.0 supports the following features:

- Visibility into inter-VM communication via NetFlow, sFlow(R), IPFIX, SPAN, RSPAN, and GRE-tunneled mirrors
- LACP (IEEE 802.1AX-2008)
- Standard 802.1Q VLAN model with trunking
- Multi-cast snooping
- IETF Auto-Attach SPBM and rudimentary required LLDP support
- BFD and 802.1ag link monitoring
- STP (IEEE 802.1D-1998) and RSTP (IEEE 802.1D-2004)
- Fine-grained QoS control
- Support for HFSC qdisc
- Per VM interface traffic policing
- NIC bonding with source-MAC load balancing, active backup, and L4 hashing
- OpenFlow protocol support (including many extensions for virtualization)
- IPv6 support

- Multiple tunneling protocols (GRE, VXLAN, STT, and Geneve, with IPsec support)
- Remote configuration protocol with C and Python bindings
- Kernel and user-space forwarding engine options
- Multi-table forwarding pipeline with flow-caching engine
- Forwarding layer abstraction to ease porting to new software and hardware platforms

2.2.2.3 Open Virtual Switch Internal Modules

The included Linux kernel module supports Linux 3.10 and up. OVS can also operate entirely in userspace without assistance from a kernel module. This userspace implementation should be easier to port than the kernel-based switch. OVS in userspace can access Linux or DPDK [83] devices. As shown in Figure 2.7, some of main components are described as follows:

- *ovs-vswitchd*, a daemon that implements the switch, along with a companion Linux kernel module for flow-based switching. It talks to the kernel module through *netlink* protocol. *ovs-vswitchd* saves and changes the switch configuration into a database and talks to *ovsdb-server* that manages *ovsdb*.

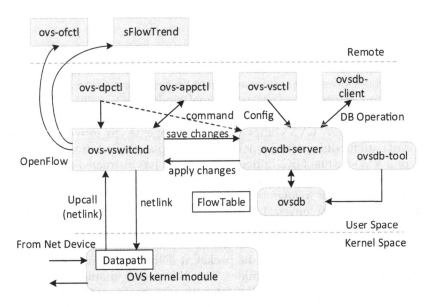

FIGURE 2.7
Open virtual switch internal modules.

- *ovsdb-server*, a lightweight database server that ovs-vswitchd queries to obtain its configuration.
- *ovs-dpctl*, the OVS datapath management utility. It configures the switch kernel module.
- *ovs-vsctl*, a utility for querying and updating the configuration of ovs-vswitchd. It manages the switch through interaction with ovsdb-server.
- *ovs-appctl*, a utility that sends commands to running Open vSwitch daemons.
- *ovsdb* persists the data across reboots; it configures ovs-vswitchd.
- *OVS kernel module* is designed to be fast and simple, and implements tunnels and caches flows. It handles switching and tunneling without knowledge of Openflow. If a flow is found, actions are executed, otherwise the flow is passed to the userspace.

Open vSwitch also provides some tools:

- *ovs-ofctl*, a management utility for querying and controlling Open-Flow switches and controllers.
- *ovsdb-tool*, a command line tool to manage database.
- *ovs-pki*, a utility for creating and managing the public-key infrastructure.

2.2.2.4 Packet Processing in OVS

In OVS, two major components direct packet forwarding [215]. The first, and larger, component is ovs-vswitchd, a userspace daemon that is essentially the same from one operating system and operating environment to another. The other major component, a datapath kernel module, is usually written specially for the host operating system for performance. Figure 2.8 depicts how the two main OVS components work together to forward packets. The datapath module in the kernel receives the packets first, from a physical NIC or a VM's virtual NIC. Either ovs-vswitchd has instructed the datapath how to handle packets of this type, or it has not. In the former case, the datapath module simply follows the instructions, called actions, given by ovs-vswitchd, which list physical ports or tunnels on which to transmit the packet. Actions may also specify packet modifications, packet sampling, or instructions to drop the packet. In the other case, where the datapath has not been told what to do with the packet, it delivers it to ovs-vswitchd. In userspace, ovs-vswitchd determines how the packet should be handled, then it passes the packet back to the datapath with the desired handling. Usually, ovs-vswitchd also tells the datapath to cache the actions, for handling similar future packets.

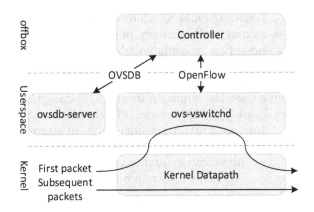

FIGURE 2.8
Packet forwarding in OVS.

OVS is commonly used as a SDN switch, and the main way to control forwarding is OpenFlow [259]. Through a simple binary protocol, OpenFlow allows a controller to add, remove, update, monitor, and obtain statistics on flow tables and their flows, as well as to divert selected packets to the controller and to inject packets from the controller into the switch. In OVS, ovs-vswitchd receives OpenFlow flow tables from an SDN controller, matches any packets received from the datapath module against these OpenFlow tables, gathers the actions applied, and finally caches the result in the kernel datapath. This allows the datapath module to remain unaware of the particulars of the OpenFlow wire protocol, further simplifying it. From the OpenFlow controllers' point of view, the caching and separation into user and kernel components are invisible implementation details: in the controllers' view, each packet visits a series of OpenFlow flow tables and the switch finds the highest-priority flow whose conditions are satisfied by the packet and executes its OpenFlow actions.

The flow programming model of OVS largely determines the use cases it can support and to this end, OVS has many extensions to standard OpenFlow to accommodate network virtualization.

2.3 Tunneling Protocols and Virtual Private Networks

The two most common forms of network virtualization are protocol-based virtual networks (such as VLANs, VPNs, and VPLSs) and virtual networks that are based on virtual devices (such as the networks connecting VMs inside a Hypervisor). Several popular virtual networking protocols are presented as follows:

- *L2TP (Layer 2 Tunneling Protocol)* is a tunneling protocol used to support VPNs or as part of the delivery of services by ISPs. It does not provide any encryption or confidentiality by itself. Rather, it relies on an encryption protocol that it passes within the tunnel to provide privacy. The entire L2TP packet, including payload and L2TP header, is sent within a User Datagram Protocol (UDP) datagram. It is common to carry Point-to-Point Protocol (PPP) sessions within an L2TP tunnel. L2TP does not provide confidentiality or strong authentication by itself. IPsec is often used to secure L2TP packets by providing confidentiality, authentication and integrity. The combination of these two protocols is generally known as L2TP/IPsec.

- *PPP (Point-to-Point Protocol)* is a data link (layer 2) protocol used to establish a direct connection between two nodes. It connects two routers directly without any host or any other networking device in between. It can provide connection authentication, transmission encryption, and compression. PPP is used over many types of physical networks including serial cable, phone line, trunk line, cellular telephone, specialized radio links, and fiber optic links such as SONET. PPP is also used over Internet access connections. Internet Service Providers (ISPs) have used PPP for customer dial-up access to the Internet, since IP packets cannot be transmitted over a modem line on their own, without some data link protocol.

- *VLAN (Virtual Local Area Network)* is any broadcast domain that is partitioned and isolated in a computer network at the data link layer. VLANs allow network administrators to group hosts together even if the hosts are not on the same network switch. This can greatly simplify network design and deployment, because VLAN membership can be configured through software. To subdivide a network into virtual LANs, one configures network equipment. Simpler equipment can partition based on physical ports, MAC addresses, or IP addresses. More sophisticated switching devices can mark frames through VLAN tagging, so that a single interconnect (trunk) may be used to transport data for multiple VLANs.

- *VXLAN (Virtual Extensible LAN)* is a network virtualization technology that attempts to improve the scalability problems associated with large cloud computing deployments. It uses a VLAN-like encapsulation technique to encapsulate layer 2 Ethernet frames within layer 4 UDP packets, using 4789 as the default IANA-assigned destination UDP port number. VXLAN endpoints, which terminate VXLAN tunnels and may be both virtual or physical switch ports, are known as VXLAN Tunnel Endpoints (VTEPs). It is an alternative of Generic Routing Encapsulation (GRE) protocol in cloud system to build private networks as layer 2 tunnels.

- *Generic Routing Encapsulation (GRE)* is a communication protocol used to establish a direct, point-to-point connection between network nodes. Being a simple and effective method of transporting data over a public network, such as the Internet, GRE lets two peers share data they will not be able to share over the public network itself. GRE encapsulates data packets and redirects them to a device that de-encapsulates them and routes them to their final destination. This allows the source and destination switches to operate as if they have a virtual point-to-point connection with each other (because the outer header applied by GRE is transparent to the encapsulated payload packet). For example, GRE tunnels allow routing protocols such as Routing Information Protocol (RIP) and Open Shortest Path First (OSPF) to forward data packets from one switch to another switch across the Internet. In addition, GRE tunnels can encapsulate multi-cast data streams for transmission over the Internet.

- *SSL (Secure Socket Layer)* is a standard security technology for establishing an encrypted link between a server and a client - typically a web server (website) and a browser, or a mail server and a mail client (e.g., Outlook) by encrypting data above the transport layer. The SSL protocol has always been used to encrypt and secure transmitted data. For example, all browsers have the capability to interact with secured web servers using the SSL protocol. However, the browser and the server need what is called an SSL Certificate to be able to establish a secure connection, where SSL Certificates are constructed based on a key pair, a public and a private key, and a certificate that contains a public key digital signed by using a trusted third party's private key. A client can use the server's certificate to establish an encrypted connection.

- *IPSec* is a network protocol suite that authenticates and encrypts the packets of data sent over a network at the IP layer. IPsec includes protocols for establishing mutual authentication between agents at the beginning of the session and negotiation of cryptographic keys for use during the session. IPsec can protect data flows between a pair of hosts (host-to-host), between a pair of security gateways (network-to-network), or between a security gateway and a host (network-to-host).

In the following subsections, we will focus on three tunneling protocols, VLAN, VXLAN, and GRE, which have been widely used in virtual networking environments.

2.3.1 VLAN

Virtual Local Area Networks (VLANs) divide a single existing physical network into multiple logical networks. Thereby, each VLAN forms its own broadcast domain. Communication between two different VLANs is only possible through a router that has been connected to both VLANs. VLANs behave as if they had been constructed using switches that are independent of each other.

2.3.1.1 Types of VLANs

In principle, there are two approaches to implementing VLANs:

- Port-based VLANs (untagged), and
- tagged VLANs.

Port-based VLANs

With regard to port-based VLANs, a single physical switch X is simply divided into multiple logical switches. The example presented in Figure 2.9 divides a ten-port physical switch into two logical switches, which are shown in Table 2.2.

All of the computers (A-H) have been connected to one physical switch; however, only the following computers can communicate with each other due to ports configured into two logical LANs: VLAN 1 and VLAN 2:

- VLAN 1: A, B, C, and D;
- VLAN 2: E, F, G, and H.

If we want to move two computers from each VLAN to a neighboring room's switch Y, and keep the same VLAN configuration, we can connect two VLANs using two physical cables as shown in Figure 2.10.

As shown in Table 2.3, in this setup, one cable is required to connect from switch X port 5 to switch Y port 1, and another cable is required to connect from switch X port 10 to switch Y port 6. In this way, two VLANs can span through two physical switches. For some switches it is necessary to set the

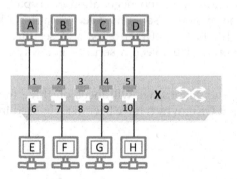

FIGURE 2.9
Port-based VLANs.

TABLE 2.2

Port-Based VLAN Assignment for Switch X

Port	VLAN ID	Connected Device
1	1	A
2		B
3		not used
4		C
5		D
6	2	E
7		F
8		not used
9		G
10		H

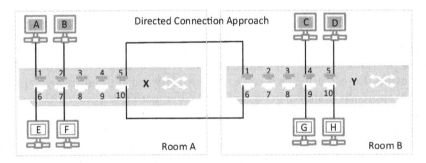

FIGURE 2.10

Port-based VLANs over multiple physical switches.

TABLE 2.3

Port-Based VLAN Assignment for Switch X and Switch Y

Port	VLAN ID	Connected Device	Port	VLAN ID	Connected Device
1	1	A	1	1	Direct connection
2		B	2		not used
3		not used	3		not used
4		not used	4		C
5		Direct connection	5		D
6	2	E	6	2	Direct connection
7		F	7		not used
8		not used	8		not used
9		not used	9		G
10		Direct connection	10		H

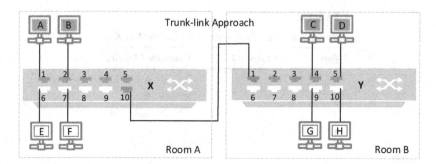

FIGURE 2.11
Tagged VLANs using a trunk link.

PVID (Port VLAN ID) on untagged ports in addition to the VLAN ID of the port. This specifies which VLAN any untagged frames should be assigned to when they are received on this untagged port. The PVID should therefore match the configured VLAN ID of the untagged port.

Tagged VLANs

In the previous example, we can reduce the physical connections from using two cables to just using one cable. This is achieved through using tagged VLAN approach as shown in Figure 2.11. With regard to tagged VLANs, multiple VLANs can be used through a single switch port that is called trunk port. Tags containing the respective VLAN identifiers indicating the VLAN to which the frame belongs are attached to the individual Ethernet frames. If both switches understand the operation of tagged VLANs in the example, the reciprocal connection can be accomplished using one single cable. The port assignment in tag-based VLANs is presented in Table 2.4.

TABLE 2.4

Tag-Based VLAN Assignment for Switch X and Switch Y

Port	VLAN ID	Connected Device	Port	VLAN ID	Connected Device
1	1	A	1	1	Trunk port
2		B	2		not used
3		not used	3		not used
4		not used	4		C
5		not used	5		D
6	2	E	6	2	not used
7		F	7		not used
8		not used	8		not used
9		not used	9		G
10		Trunk port	10		H

2.3.1.2 IEEE 802.1Q

IEEE 802.1Q is a protocol for carrying VLAN traffic on an Ethernet. A VLAN is a type of local area network that does not have its own dedicated physical infrastructure, but instead uses another LAN to carry its traffic. The traffic is encapsulated so that a number of logically separate VLANs can be carried by the same physical LAN.

VLANs are used whenever there is a need for traffic to be segregated at the link layer. For example, on Internet Protocol networks it is considered good practice to use a separate VLAN for each IP subnet. Reasons for doing this include:

- preventing a machine assigned to one subnet from joining a different one by changing its IP address; and
- avoiding the need for hosts to process broadcast traffic originating from other subnets.

Tagging

As shown Figure 2.12, in 802.1Q VLAN frames are distinguished from ordinary Ethernet frames by the insertion of a 4-byte VLAN tag into the Ethernet header. It is placed between the source MAC and the EtherType fields:

The first two bytes of the tag contain the Tag Protocol Identifier (TPID), which is defined to be equal to 0x8100. Since it is located where the EtherType would appear in an ordinary Ethernet frame, tagged frames appear to have an EtherType of 0x8100. The remaining two bytes contain the TCI (tag control information), of which 12 bits correspond to the VLAN Identifier (VID), and 4 bits contain metadata used for quality of service management.

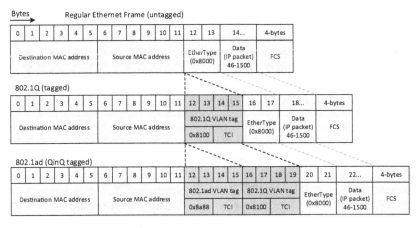

FIGURE 2.12
Ethernet Frame, 802.1Q (VLAN), and 802.1ad (QinQ).

VLAN numbering

Each 802.1Q VLAN is identified by a 12-bit integer called a VID (VLAN Identifier) in the range 1 to 4094 inclusive. The values 0 and 4095 are reserved and should not be used. The first VLAN, with a VID of 1, is the default VLAN to which ports are presumed to belong if they have not been otherwise configured. It is considered good practice to move traffic off the default VLAN where possible, in order to minimize the extent to which an unconfigured switch port would give access to the network. This does not mean that a network, which uses the default VLAN is necessarily insecure, and vacating it may not be feasible since some devices have functions that are hardwired to a VID of 1. However, if we have the opportunity to use a different VLAN, then that is usually preferable.

The remaining values have no special status and can be used freely but be aware that many network devices place a limit on the number of VLANs that can be configured so it will not necessarily be feasible to make use of all 4094 possible VIDs.

Trunk and access ports

There are two ways in which a machine can be connected to a switch carrying 802.1Q VLAN traffic:

- via an access port, where VLAN support is handled by the switch (so the machine sees ordinary, untagged Ethernet frames); or
- via a trunk port, where VLAN support is handled by the attached machine.

It is also possible to operate a switch port in a hybrid mode, where it acts as an access port for one VLAN and a trunk port for others. Thus, the attached Ethernet segment can carry a mixture of tagged and untagged frames. This is not recommended due to the potential for VLAN hopping.

Port configuration

The 802.1Q standard does not itself make any formal distinction between trunk and access ports. Instead, it allows the manner in which each VLAN is handled to be configured separately. In Table 2.5, it shows each port to be in one of three states for a given VLAN:

TABLE 2.5

Port Configurations of 802.1Q Frame

State	Ingress	Egress
tag	allowed	allowed, will be tagged
untag	allowed	allowed, will not be tagged
non-member	prohibited	prohibited

In addition to this, each port has a PVID (Port VLAN ID) which specifies which VLAN any untagged frames should be assigned to. It may also be possible to specify which frame types are acceptable for ingress (tagged, untagged or both).

This method of configuration provides a lot of flexibility, but be aware that just because a configuration is possible does not mean that it is useful or safe. For all but the most unusual purposes you should configure each port so that it is either an access port or a trunk port:

- For an access port, there should be exactly one untagged VLAN (the one to be made accessible) and no tagged VLANs. The PVID should be set to match the untagged VLAN. If there is a choice, then the port should admit only untagged frames.

- For a trunk port, there may be any number of tagged VLANs, but no untagged VLANs. Ideally, the PVID should be set to a VLAN that does not carry any legitimate traffic, but this is not essential. If there is a choice, then the port should admit only VLAN-tagged frames.

Effect on the MTU

The Maximum Transmission Unit (MTU) of a network interface is the size of the largest block of data that can be sent as a single unit. The standard Ethernet MTU is 1500 bytes at the network layer or 1518 bytes at the link layer, the difference is due to the 14-byte header and 4-byte frame check sequence that enclose the payload of an Ethernet frame.

On a VLAN trunk the need for each frame to be tagged adds a further 4 bytes of link-layer framing. This can be accommodated either by increasing the link-layer MTU or by reducing the network-layer MTU:

- To use the standard network-layer MTU of 1500 bytes, the equipment must support a link-layer MTU of at least 1522 bytes.

- If the link-layer MTU were limited to the standard value of 1518 bytes, then the network-layer MTU would need to be reduced to 1496 bytes to compensate.

Devices with explicit VLAN support are supposed to accommodate a link-layer MTU of at least 1522 bytes, but if you are using generic hardware then it may be necessary to accept a lower value. All devices on a given IP subnet must use the same network-layer MTU, so if you intend to deviate from the standard value of 1500 bytes then you will need to configure all affected machines. Similar considerations apply when using jumbo frames. The link layer MTU is then much larger, but so is the potential payload, so allowance must still be made.

VLAN stacking

Because an 802.1Q VLAN can carry arbitrary Ethernet traffic, it is in principle feasible to nest one VLAN within another. Possible reasons for doing this include:

- carrying more than 4094 separate VLANs on one physical bearer, simplifying the configuration of backbone switches, or
- allowing customer and service-provider VLANs to be administered independently of each other.

A basic 802.1Q-compatible switch cannot be used to add further tags to an already tagged frame, but there is an amendment to the standard called IEEE 802.1ad (also known as QinQ) which adds this capability. Note that VLAN stacking exacerbates the effect on the MTU, as each extra level adds a further 4 bytes of link-layer framing.

The QinQ technology is also called VLAN dot1q tunnel, 802.1Q tunnel, and VLAN Stacking technology. The standard comes from IEEE 802.1ad and it is the expansion of the 802.1Q protocol. QinQ adds one layer of 802.1Q tag (VLAN Tag) based on the original 802.1Q packet head. With the double layers of tags, the VLAN quantity is increased to 802.1Q.

QinQ encapsulates the private network VLAN Tag of the user in the public (service provider) network VLAN Tag to make the packet with double layers of VLAN Tags cross the backbone network (public network) of the operator. In the public network, the packet is passed according to the out-layer of VLAN Tag (that is the public network VLAN Tag) and the private network VLAN Tag of the user is shielded.

The formats of the common 802.1Q packet with one layer of VLAN Tag and the QinQ packet with two layers of VLAN Tags are shown in Figure 2.12. Two layers of VLAN tags can support 4Kx4K VLANs, meeting most of requirements ISP and overcoming limitation of VLANs.

In addition to increase the number VLANs, QinQ also provides one simple L2 VPN tunnel for the user, and it does not need the supporting of the protocol and signaling, which can be established by using static configuration. QinQ is divided into two kinds, including basic QinQ and selective QinQ.

- *Basic QinQ*: When receiving the packet, the QinQ port adds the VLAN Tag of the default VLAN of the port to the packet no matter whether the packet has the VLAN Tag. Before the packet is forwarded out from the QinQ port, delete the out-layer of Tag and then forward it. The disadvantage of the method is that the encapsulated out-layer of VLAN cannot be selected according to the VLAN Tag of the packet.
- *Selective QinQ*: The selective QinQ solves the disadvantage of the basic QinQ. When receiving the packet, the QinQ port adds the specified-out layer of VLAN Tag to the packet according to the VLAN Tag of the packet. If the encapsulated out-layer of VLAN Tag is not specified, add the VLAN Tag of the default VLAN of the port to the packet.

QinQ can be used to address three major network setup requirements: (1) it can shield the VLAN ID of the user, so as to save the public network VLAN ID resource of the service provider, (2) the user can plan the private network

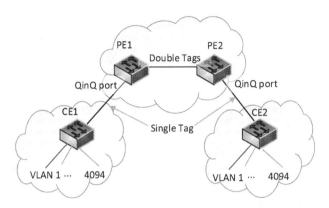

FIGURE 2.13
A QinQ example.

VLAN ID, avoiding the conflict with the public network and other user VLAN IDs, and (3) it provides the simple L2 VLAN solution.

To illustrate how QinQ works, an example is presented in Figure 2.13. The upstream packet of the CE1 switch carries one layer of VLAN tag. The packet reaches the QinQ port of the PE1 switch. According to the configuration of the QinQ port, add one out-layer of VLAN Tag to the packet. The packet with two layers of VLAN tags is forwarded to PE2 via the public network. On the QinQ port of PE2, the out-layer of VLAN Tag is deleted, and the packet recovers to have one layer of VLAN Tag and is forwarded to CE2.

VLAN hopping

802.1Q VLANs can be expected to provide a degree of isolation that is almost as good as would be provided by separate physical networks. This isolation is not complete because there will usually be competition for shared network bandwidth, but if the infrastructure has been securely configured then no traffic should be able to enter a VLAN unless it has been deliberately routed or bridged there by one of the connected hosts.

Special care is needed to achieve this state of affairs, because some configurations can be exploited to inject unauthorized traffic into a VLAN. This practice is known as VLAN hopping and is usually accomplished by:

- double tagging, or
- somehow persuading a switch to reconfigure an access port as a trunk port.

One method than can sometimes be used to hop between VLANs is to construct a frame with two tags. If this traverses a VLAN-aware switch that has been poorly configured, then it may be possible to forward the frame onto a VLAN trunk with its outer tag removed but the inner tag intact and exposed. There are two conditions that must be satisfied for this attack to be feasible:

- The egress port on the switch must operate in the hybrid mode described above, where traffic belonging to one of the possible VLANs is forwarded in untagged form.
- The ingress port must allow access to that VLAN by means of a tagged frame.

The inner tag should match the VLAN you want to hop to. The outer tag should match the VLAN that is untagged on the egress port. An effective way to defend against this technique is to ensure that the conditions described above do not arise. Specifically, you should ensure that every active port is either:

- a trunk port which tags all outbound frames for all VLANs, or
- an access port which does not tag any frames.

As an additional protection, some switches may prohibit the ingress of tagged frames on ports that are supposed to be access ports. However, we should be aware that whether frames are tagged or untagged on egress normally has no bearing on which frame types are acceptable for ingress.

2.3.2 Virtual Extensible LAN

Many enterprise and service provider customers are building private or public clouds. Intrinsic to cloud computing is the presence of multiple tenants with numerous applications using the on-demand cloud infrastructure. Each of these tenants and applications needs to be logically isolated from the others, even at the network level. For example, a three-tier application can have multiple virtual machines in each tier and requires logically isolated networks between these tiers. Traditional network isolation techniques such as IEEE 802.1Q VLAN provide 4096 LAN segments (through a 12-bit VLAN identifier) and may not provide enough segments for large cloud deployments. Moreover, in a distributed cloud networking environment, it is common to establish a VLAN across geographically remote datacenters.

To address these issues, several Cisco, VMware, Citrix, and Red Hat worked together to design and promote Virtual Extensible LAN (VXLAN) solutions to address new requirements for scalable LAN segmentation and for transport of virtual machines across a broader network range. VXLAN defines a 24-bit LAN segment identifier that provides segmentation at cloud scale. In addition, VXLAN provides an architecture that customers can use to expand their cloud deployments with repeatable pods in different Layer 2 domains. VXLAN can also enable migration of virtual machines between servers across Layer 3 networks.

2.3.2.1 VXLAN Design Requirements and Challenges

Cloud Computing Demands More Logical Networks

An infrastructure for a service cloud computing environment can have a large number of tenants, each with its own applications. In fact, each tenant requires a logical network isolated from all other tenants. Furthermore, each application from a tenant also requires its own logical network, to isolate it from other applications. To provide instant provisioning, cloud management tools, such as VMware vCloud Director, clone the application's virtual machines, including the virtual machine's network addresses, that demands a logical network for each instance of the application.

Challenges of Existing Network Isolation Techniques

The VLAN has been the traditional mechanism for providing logical network isolation. Because of the ubiquity of the IEEE 802.1Q standard, numerous switches and tools are available that provide robust network troubleshooting and monitoring capabilities, enabling mission-critical applications to depend on the network. The IEEE 802.1Q standard specifies a 12-bit VLAN identifier, which limits the scalability of cloud networks beyond 4K VLANs. Some in the industry have proposed incorporation of a longer logical network identifier in a MAC-in-MAC or MAC in Generic Route Encapsulation (MAC-in-GRE) encapsulation as a way to expand scalability. However, in a data center network port channels are commonly used, where a port channel is an aggregation of multiple physical interfaces that create a logical interface. Up to eight individual active links can be bundled into a port channel to provide increased bandwidth and redundancy. In addition, port channeling also load-balancing traffic across these physical interfaces. Existing layer-2 virtual networking solutions transport network packets inefficiently because they cannot make use of all the links in a port channel, which is typically implemented in data center networks.

2.3.2.2 VXLAN Frame

VXLAN meets these challenges with a MAC in User Datagram Protocol (MAC-in-UDP) encapsulation technique and a 24-bit segment identifier in the form of a VXLAN ID (Figure 2.14). The larger VXLAN ID allows LAN segments to scale to 16 million in a cloud network. In addition, the UDP

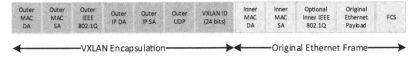

FIGURE 2.14
A VXLAN frame.

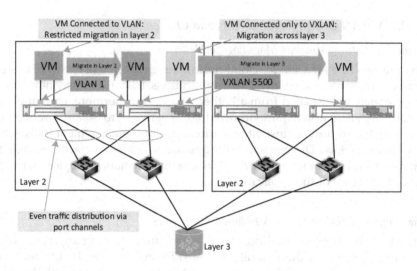

FIGURE 2.15
A VXLAN example with port channels and layer-3 migration.

encapsulation allows each LAN segment to be extended across Layer 3 and helps ensure even distribution of traffic across port channel links.

VXLAN uses an IP multi-cast network to send broadcast, multi-cast, and unknown unicast flood frames. When a VM joins a VXLAN segment, the server joins a multi-cast group. Broadcast traffic from the VM is encapsulated and is sent using multi-cast to all the servers in the same multi-cast group. Subsequent unicast packets are encapsulated and unicast directly to the destination server without multi-cast. In effect, traditional switching takes place within each VXLAN segment.

In the example presented in Figure 2.15, VXLAN solution provides the following capabilities:

- Logical networks can be extended among virtual machines placed in different Layer 2 domains, e.g., VXLAN 5500.
- Flexible, scalable cloud architecture enables addition of new server capacity over Layer 3 networks and accommodates elastic cloud workloads, i.e., two geo-separated domains can be connected in Layer 2.
- If a VM is connected only through VXLAN, then it can migrate across the Layer 3 network.
- If a VM is connected to a VLAN, then it is restricted to migration within the Layer 2 domain. Note that a VM on a VXLAN segment needs to be connected to a VLAN network in order to interact with external networks.

2.3.3 Generic Routing Encapsulation

Generic Routing Encapsulation (GRE) is a communication protocol used to establish a direct, point-to-point connection between network nodes. Being a simple and effective method of transporting data over a public network, such as the Internet, GRE lets two peers share data they are able to share over the public network itself.

GRE in its simplest form provides a way to encapsulate any network layer protocol over any other network layer protocol. For example, GRE can be used to transport IP packets with private IP addresses over the public Internet to a remote private IP network. GRE allows routers to act as if they have a virtual point-to-point connection to each other. GRE tunnels allow routing protocols (like RIP and OSPF) to be forwarded to another router across the Internet. In addition, GRE tunnels can encapsulate multi-cast data streams for transmission over the Internet.

GRE tunneling is accomplished by creating routable tunnel endpoints that operate on top of existing physical and/or other logical endpoints. By design, GRE tunnels connect A to B and provide a clear data path between them. To protect the payload, usually, IPSec must be used on the GRE tunnels to ensure data security. Data is routed by the system to the GRE endpoint using routes established in the route table. Once a data packet is received by the GRE endpoint, it is encapsulated in a GRE header and routed again using the endpoint configuration (destination address of the tunnel); therefore, each data packet traveling over the GRE tunnel gets routed through the system twice. An IP packet routed through a GRE tunnel can be viewed in Figure 2.16.

2.3.3.1 GRE Header

Figure 2.17 presents the GRE packet header format. The GRE packet header varies in length (depending on options) and can be 32 to 160 bits long. A standard GRE header has the form:

Delivery header	GRE header	Original packet

FIGURE 2.16
GRE packet encapsulation.

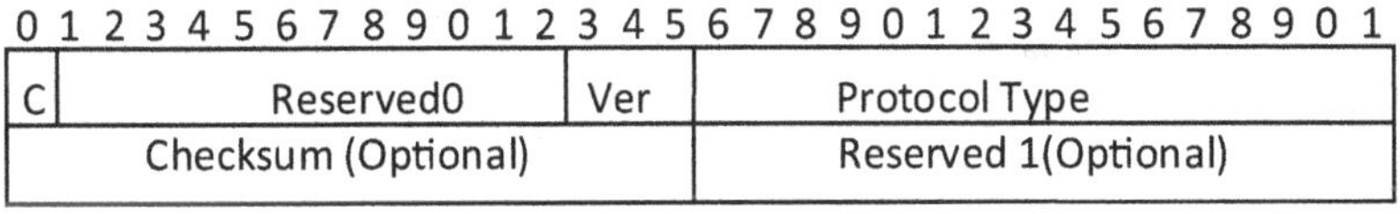

0 1 2 3 4 5 6 7 8 9 0 1 2 3 4 5 6 7 8 9 0 1 2 3 4 5 6 7 8 9 0 1

C	Reserved0	Ver	Protocol Type
Checksum (Optional)			Reserved 1(Optional)

FIGURE 2.17
GRE packet format.

- Checksum present (C, Bit 0) specifies whether there is a checksum in the GRE header. When it is set to 1, the checksum and reserved1 fields (bits 32 to 63) must contain appropriate information. When set to 0, the checksum and reserved1 fields are not necessary.
- Reserved0 (Bits 1 to 12) contains several bits to indicate the presence of optional fields in the header (per RFC2890).
- Version (Ver, Bits 13 to 15) must contain 000 for GRE.
- Protocol Type (2 octets) specifies the protocol type of the original payload packet as defined in RFC1700. For SROS GRE applications, this will always be 0x800 (IP).
- Checksum (2 octets), when the checksum present bit is set to 1, the checksum octets contain the IP checksum of the GRE header and payload packet.
- Reserved1 (2 octets) is reserved for future use. It is only presented when the checksum present bit is set to 1.

2.3.3.2 GRE Packet Flow

Tracing a packet destined for a network available through the GRE tunnel will provide a better functional understanding of how the GRE implementation works. Using the network diagram in Figure 2.18 as our example, trace a packet from a node on the 192.168.1.0 network (192.168.1.10) to a node on the 192.168.2.0 network (192.168.2.15). The step-by-step tunnel establishment procedure is presented in Table 2.6.

2.4 Virtual Routing and Forwarding

Virtualization is a technique for hiding the physical characteristics of computing resources from the way in which other systems, applications, or end users

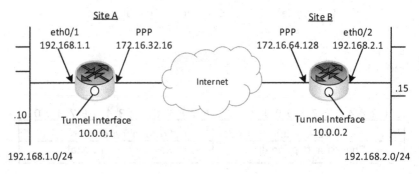

FIGURE 2.18
GRE tunnel establishment example.

TABLE 2.6

GRE Tunnel Establishment

Step	Description
1	Packet originates from 192.168.1.10.
2	Packet received by the router at Site A on eth 0/1 (192.168.1.1).
3	Checks routing table for routing information (destination 192.168.2.15) and determines the destination network is available through the tunnel interface.
4	Packet is encased in GRE header with source IP (172.16.32.16) and destination IP (172.30.64.128) and routed back through the route stack.
5	Checks routing table for routing information for destination IP 172.30.64.128 and routes the packet out the WAN interface.
6	The router at Site B receives the packet on the PPP interface (172.30.64.128). The system recognizes that there is a GRE header on the packet and sends it to the tunnel interface associated with the source and destination IP addresses (172.16.32.16 and 172.30.64.128).
7	GRE header is stripped and the packet is routed through the route stack with a destination IP of 192.168.2.15.
8	Packet is routed out eth 0/2 for delivery.

interact with those resources. This includes making a single physical resource (such as a server, an operating system, an application, storage device, or network) appear to function as multiple logical resources; or it can include making multiple physical resources (such as storage devices or servers) appear as a single logical resource. Virtual networks is a generic term that uses many different technologies to provide virtualization. Fundamentally, virtual networks all provide a mechanism to deploy what looks and operates like multiple networks and are actually all using the same hardware and physical connectivity.

If we needed IP networks that are isolated as they were used by different companies, departments, or organizations, we would normally deploy multiple IP networks made up of separate physical routers that were not connected to each other. They may still be using a shared Layer 2 or Layer 1 infrastructure; however, at Layer 3 they are not connected and do not form a network. Network virtualization allows a single physical router to have multiple route tables. The global table contains all IP interfaces that are not part of a specific virtual network and route tables are for each unique virtual network assigned to an IP interface. In its basic form, this allows an Ethernet 0/0 IP interface to be in virtual network 10 and Ethernet 0/1 IP interface to be in virtual network 20. Packets arriving on Ethernet 0/0 are only forwarded to other interfaces in virtual network 10 and do not use Ethernet 0/1, because it is not in its virtual network: virtual network 10 has no routing knowledge of other virtual networks. Additional virtualization can be provided by allowing multiple virtual networks per physical connection. This is enabled by using Layer 2 logical connections. For an Ethernet physical port, the use of multiple virtual LANs (VLANs) allows each VLAN to use a different virtual network.

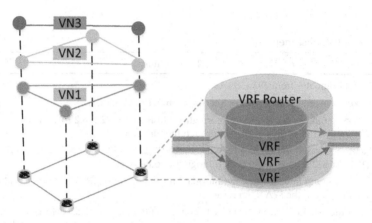

FIGURE 2.19
Virtual route forwarding (VRF).

Virtual Route Forwarding (VRF) is a technique which creates multiple virtual networks within a single network entity (as shown in Figure 2.19). In a single network component, multiple VRF resources create the isolation between virtual networks. The goal of a VRF is to build a separate routing table that is independent of the main one. VRFs are the same methods of network isolation/virtualization as VLANs, VLANs are used at the Layer-2 and VRFs are Layer-3 tools.

VLAN is the existing virtualization technique at Layer 2 to separate different types of traffic, e.g., voice networks from data networks. However, in a case where a soft-phone from a PC (in a data VLAN) wants to talk to a hard-phone in a voice VLAN, using a global routing table, it creates a potential security threat since the voice VLAN is now visible from the data network.

Security can be improved by deploying virtualization at the network level. VRF technology can be used with the rules and policies so that each VRF network achieves the expected level of security. The system allows only the soft-phone application to talk to the hard-phone in a voice VRF, or allow different soft phones within data VRF resources to talk to each other; however, do not allow other applications to communicate with each other.

Summary

Understanding virtual networking concept is essential for understanding cutting-edge programmable networking technologies. This chapter is one of the important chapters of this book for readers to build the knowledge foundation to understand modern computer networking solutions, especially later when we describe network functions virtualization and software-defined networking.

Various virtual networking solutions also help us build effective security management systems such as VPN technologies which are good examples to provide network-level isolations to protect network traffic. Moreover, virtual networks provide us an approach to change the management or protection granularity of networks, which later we refer to as network micro-segmentation in Chapter 6. Based on the provisioned virtual networking resource, we can provide customized and adaptive security provisioning such as monitoring, detection, and recovery security appliances via network functions virtualization solutions. Interested readers can jump to Chapter 6 if you have sufficient background of SDN and NFV.

3

SDN and NFV

How does underlying technology change when confronted with changing use patterns or changing user requirements? In most cases, the driver is purely economical. If the investment in a technology is large enough, it is stretched to meet new needs. In this chapter, we examine two new paradigms that helped "stretch" the traditional IP infrastructure in the information age. First amongst these is Network Functions Virtualization (NFV) which, as the name suggests, moved network functions from stand-alone appliances to software running on any server, reducing the time-to-market for products. The second is a more clean slate approach using SDN, wherein separation of control and data plane leads to a highly agile and dynamic environment. These two technologies, along with some benefits, challenges and use cases are described in detail next.

3.1 Introduction

The current era of human history, often called the Information Age, is characterized by the ubiquity of information. Since its advent in the 1960s, pervasiveness of Internet has resulted in nearly 40% of the global population using this technology to revolutionize the way we do business, socialize, gain knowledge, and entertain ourselves [15]. The explosion of information has resulted in a realization of how, despite their prevalence, the traditional IP-based networks are inflexible and difficult to manage. This, in short, is the motivation for Network Functions Virtualization (NFV) and SDN. While they are closely related and often coexist, NFV and SDN are distinct approaches to introducing malleability to the infrastructure that forms the bedrock of the information age.

As the name suggests, NFV is the process of moving network services like firewalls, Network Address Translation (NAT), load balancing, Intrusion Detection System (IDS), etc. from dedicated physical hardware into services on a virtualized environment. The movement to virtualize is not limited to network functions, and is part of the broader shift in computing industry to better employ the immense gains in hardware capabilities. Generally, NFV does not introduce changes to existing protocols, and can be implemented incrementally. SDN, on the other hand, seeks to introduce flexibility, dynamism and automation into the deployment and management of network objects by decoupling and centralizing the network intelligence from the packet-forwarding process. While it is possible to virtualize network functions without using SDN concepts and to realize an SDN that does not use NFV, there are inherent synergies in leveraging SDN to implement and manage an NFVI; leading to SDN and NFV being considered symbiotic paradigms.

3.2 Network Functions Virtualization

NFV is the concept of moving dedicated network functions from stand-alone appliances to software running on any white box or Commercial off-the-Shelf (COTS) server. By fundamentally altering the design and deployment process that service providers have used for a generation, NFV aims to inject dynamism and cost efficiency to an antiquated domain that suffered from chronic issues.

3.2.1 Background and Motivation behind NFV

Postal, Telegraph and Telephone (PTT) service providers (carriers or telcos) were legacy government agencies [238] that evolved to become mobile wireless carriers and data service providers by becoming Internet Service Providers (ISPs). As such, the term *carriers* is now used to describe both, the classic telephone companies and pure ISPs.

By virtue of their origins, sanctions of new products and services by a carrier followed rigorous standards for reliability and quality. The endorsement *carrier grade* is still used to designate devices that pass such strict standards. Over time, such standards led to carriers buying hardware from a small list of highly specialized vendors that could satisfy them. However, this need for reliability and quality impeded product development and led to carriers falling behind competition from market disrupters like Google, Apple, Microsoft, Amazon, and Facebook in providing value-added services on the public Internet. To maintain revenue and track their growth objectives, the carriers sought to shorten product development cycles by leveraging virtualization technologies that had made their impact felt in the computing world. To accelerate progress, several European carriers banded together and created the European Telecommunications Standards Institute (ETSI). The ETSI Industry Specification Group for Network Functions Virtualization (ETSI ISG NFV), published a whitepaper titled, "Network Functions Virtualisation" [59] in October 2012 to outline the rationale, benefits, enablers and challenges for NFV with the goal to accelerate development and deployment using generic servers. As a paradigm, NFV was born.

Using the newly minted NFV framework, carriers hoped to alleviate the following concerns:

- Since large swaths of the carrier infrastructure had a few vendors, carriers were often at the mercy of the hardware manufacturers for deployment of services. The design-procure-integrate-deploy cycle that accompanies adding new hardware made operation expensive.
- Hardware installed in the late 1990s and early 2000s was rapidly reaching end of life, and replacing them en-masse was costly.
- Proving value-add services meant having to diversify the type of hardware present in the carrier environment. Increasing the variety of hardware vendors did not always ensure smooth interoperablity.
- The presence of multiple hardware products resulted in carriers having to deal with chronic complexity management of hardware, firmware, and software updates.
- Carriers had little visibility into security issues with the proprietary hardware.
- Launching new services was arduous, with carriers having the need to allocate space and power, along with integration of another hardware device into a very large network.

Since the publication of the white paper, ETSI ISG NFV has produced more detailed documentation, standard terminology, and use cases that seek to advise adopters and steer the direction of NFV though their instrumental work in collaborative projects like OPNFV [11].

3.2.2 NFV Framework

The NFV framework, thought of by ETSI, consists of three main components:

- Virtual Network Functions (VNFs) are the software implementations of network functions. VNFs are used as modules that are deployed as the building blocks of an NFV architecture.
- NFVI is the entirety of the hardware and software components that build the environment where VNFs are deployed. Processing (both virtual and physical), storage resources and virtualization software are essential parts of the NFVI. If the infrastructure is distributed, the network backbone providing connectivity between the sites is considered part of the NFVI.
- NFV Management and Orchestration (NFV-MANO) consists of all functional blocks, data repositories, reference points, and interfaces that are used for managing and orchestrating VNFs and the NFVI.

3.2.3 Benefits and Challenges of NFV

NFV has been a catalyst for major change in the telecommunications domain. As was originally intended with the publication of the whitepaper in 2012, NFV simplifies and speeds up the process of deploying new functions or applications that help the carrier provide value-add services or save costs.

Automating the orchestration and management lets an NFVI be more scalable and achieve better resource utilization. Using generic server hardware that can act as any number of network devices instead of specialized hardware that perform singular functions helps reduce both operational and capital expenditures. Further, virtualization offers carriers the ability to offer pay-as-you-go services without huge up-front investment. Avoiding proprietary hardware empowers administrators with a streamlined provisioning process.

If a service provider's customer requests a new function, for example, NFV enables the service provider to more easily add that service in the form of a VM without upgrading or buying new hardware. Consider Multimedia Messaging Service (MMS), which is used to send messages with multimedia content over a cellular network. During the early 2000s, the MMS infrastructure included at least a dozen dedicated servers that had specialized functions in addition to a multimedia gateway. Launching MMS, or increasing capacity was expensive for a telco - and hence for the user. However, with NFV, a telco can add capacity as needed with virtualized generic servers, thereby reducing costs drastically.

Being in the early stages of adoption, however, results in a few challenges facing NFV as well. Foremost among these are a lack of universally accepted standards in NFV MANO and Operational and Billing Support Systems (OSS/BSS). As with the case of the videotape format war (VHS vs. Betamax) of the late 1970s, and to a lesser extent the high definition optical disc format war (Blu-ray vs HD-DVD) of the early 2000s, carriers are hesitant to invest in a standard that is not universally accepted.

3.2.4 OPNFV

Open Platform for NFV (OPNFV) [11] is a collaborative open-source platform that seeks to develop NFV and shape its evolution. Created by the Linux Foundation in 2014, contributors now include Cisco, Juniper Networks, Dell, Hewlett-Packard, Intel, AT&T, Brocade Communications Systems, China Mobile, Ericsson, Huawei, IBM, NEC, Nokia Networks, Red Hat, etc. By bringing together upstream and downstream organizations for integrated development and testing, the OPNFV project looks to reduce the time to market for NFV products and solutions. By cooperative development, OPNFV helps with life-cycle management, introducing dynamism and failure detection in a multi-vendor environment. A case study that details the use of OPNFV to implement a virtual Customer Premise Equipment (v-CPE) is described next.

3.2.5 OpenStack

OpenStack is an open-source cloud computing platform that has high market penetration. Since its inception in 2010 as a joint project of Rackspace Hosting and NASA, OpenStack has evolved to a community project with more than 500 active contributors managed by the OpenStack Foundation [10]. It includes a collection of interoperable modules that are used to orchestrate large pools of compute, storage, and networking resources. These resources are managed either through command line tools or a web-based GUI. Further, OpenStack provides RESTful APIs to its back-end. We present an overview of the most commonly used OpenStack modules (Figure 3.1) and the interconnectivity between them before presenting a case study of an OpenStack implementation.

The modular nature of OpenStack coupled with it being a community project has led to a multitude of modules, each of which strives to achieve a clearly stated objective: help deploy OpenStack. Modules in OpenStack span the entire range of the CMMI maturity scale [127]. We limit our discussion to the more mature OpenStack modules.

In Figure 3.2, we classify modules using their functionality as the basis. Succinctly, they are as follows:

- Core modules - Modules that are considered essential to operating an OpenStack infrastructure provide basic compute, network and storage capabilities along with security provided by identity management. They fall into both the VNF and NFVI components of the NFV framework. The core modules are among the more mature modules, and the most widely deployed. When asked, *"What is the bare minimum modules I need to install to get a webserver running using an OpenStack cloud?"*, these would be the modules you install and configure. These modules are the following:

  - **Nova**, or compute module, is used to create and delete compute instances as required.

  - **Glance** synchronizes and maintains VM images across the compute cluster.

  - **Keystone** provides authentication for accessing all of OpenStack's services.

  - **Cinder** provides block storage used as storage volumes for VMs.

  - **Swift** provides object storage that is used to store large amounts of static data in a cluster.

  - **Neutron**, or networking, allows the different compute instances and storage nodes to communicate with each other.

- Management modules - Modules that enhance the MANO experience in the OpenStack installation fall into this category. While some

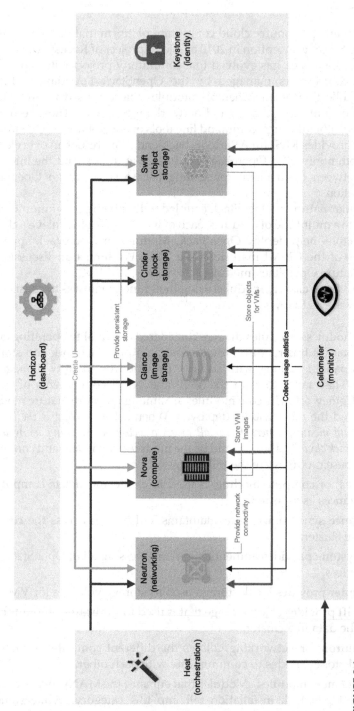

FIGURE 3.1
OpenStack module interconnectivity.

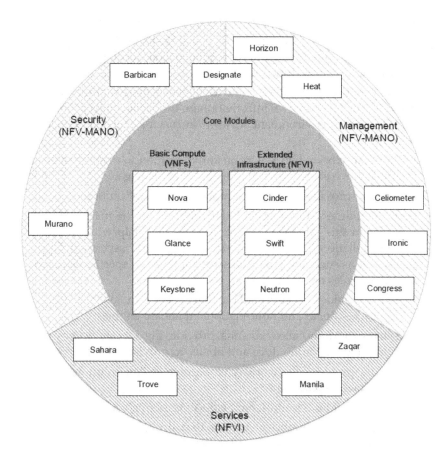

FIGURE 3.2
Categorization of modules in OpenStack.

management modules are mature, there are several projects that are
evolving rapidly.

- **Horizon** provides a GUI dashboard, and is by far the most widely
  deployed management module.
- **Heat** helps expedite orchestration of applications across multiple
  compute instances by using templates.
- **Celiometer** monitors the NFVI and helps identify bottlenecks and
  resource optimization opportunities.
- **Ironic** is a provisioning tool for baremetal installation of compute
  capabilities instead of VMs in OpenStack. The Ironic module is
  forked from the Nova baremetal driver.
- **Congress** is a policy management framework for the OpenStack
  environment.

- **Designate** is used to point applications in the OpenStack environment to a trusted DNS source. Since it seeks to enhance the security in the environment, it is often designated as straddling the management and security functions.
- Security modules - Relatively new to the scene, these modules seek to provide trusted sources of information for the applications running in the OpenStack environment to keep intrusions and malware to a minimum.
  - **Barbican** works with Keystone authentication to manage internal application security by behaving as a key manager.
  - **Murano** provides a white list repository of applications.
- Service modules - Once again, relatively low on the maturity scale, modules in this category provide services that help with an OpenStack installation. They do not address any generic issues with OpenStack, and as such, these modules are deployed in specific use cases. Some examples are as follows:
  - **Trove** provides a distributed database service and enables users to deploy relational and non-relational database engines.
  - **Sahara**, formerly called Savanna, provides big data services by providing Elastic MapReduce and ability to provision Hadoop.
  - **Manila** provides Network Area Storage (NAS) solutions for an OpenStack deployment.
  - **Zaqar** provides a multi-tenant cloud messaging service.
  - **Magnum** is an umbrella project that provides containerization assistance. This module is still in development.

3.3 Software-Defined Networks

The SDN paradigm is based on the premise that separating control of network functions from the network devices themselves (switches, routers, firewalls, etc.) can address several limitations associated with today's vertically integrated, closed and proprietary networking infrastructure. The adoption of virtualization technologies in computing, and the convergence of voice, video and data communication to IP networks fueled the need for such a shift in networking standards [77]. Figure 3.3 shows a typical network implemented using SDN in a data center environment. Four users have VMs running on the same physical host, with each VM connected to the same OVS (described in Section 3.3.7). Data frames that come from the VMs are tagged with a VLAN ID or some other ID based on the tunneling protocol in use, logically separating each of the four users. The OVS then

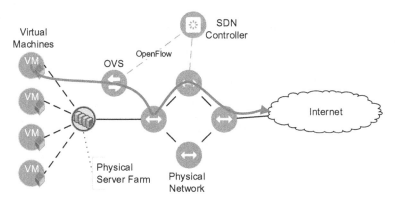

FIGURE 3.3
Typical network implemented using SDN.

uses flow rules it gets from the SDN controller to determine how to handle the traffic.

The separation of the control and data planes result in network switches becoming dumb forwarding devices, with control logic being implemented in a centralized[1] controller [185]. This not only allows the network administrators a much finer granularity of control over traffic flow, but also empowers them to respond to changing network requirements in a dynamic environment [36] in a much more effective manner. Figure 3.4 shows a simplified view of the SDN architecture.

3.3.1 Benefits and Challenges of SDN

Use of SDN has picked up steam due to the following benefits:

- The traffic patterns culminating from the adoption of cloud systems and big-data computing do not adhere to the traditional notion of a north-south network.

- Separating network control from the hardware devices eliminates the need to configure each device individually. Having a central network policy that can be dispatched to the SDN devices reduces the time-to-deploy thereby enhancing profits for the data center or service providers.

- Since control is separated from the network devices, administrators can modify the behavior of the device by pushing software updates to the device, instead of conducting fork-lift upgrades - once again enhancing profits for data center providers.

---

[1] The controller only needs to be *logically* centralized. This may be implemented in a physically centralized or distributed system [159].

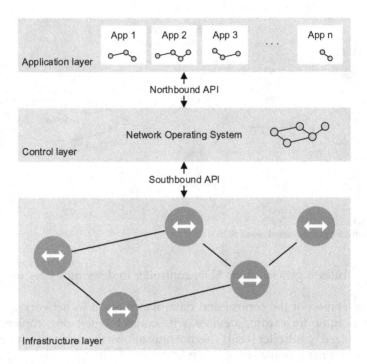

FIGURE 3.4
Abstraction in SDN.

- A singular device can handle the functionalities managed by multiple traditional network devices. For example, a single device could do switching, routing, load balancing and security functions. Further, SDN is vendor agnostic, thereby allowing providers more flexibility.
- SDN can organically provide traffic shaping and administer QoS. In current networks, provisioning different QoS levels for different applications is a highly manual process, and cannot dynamically adapt to changing network conditions [147].
- SDN provides a layer of abstraction that allows application managers and administrators to dissociate from managing the physical hardware. In addition to having access to virtual disk and memory, SDN virtualizes a Network Operating System (NOS), abstracting the physical topology of the network from the applications. As shown in Figure 3.4, several applications running on the same physical hardware could have different views of the network.

In addition to being deployed for a variety of traditional functionalities like routing, security and load balancing, SDN can be used for traffic engineering, end-to-end QoS enforcement, mobility management, data center implementation and reducing power consumption. Kreutz et al. [164] group all these applications into five categories: 1) traffic engineering; 2) mobility and wireless; 3) measurement and monitoring; 4) security; and 5) data center networking.

3.3.2 Background

Despite their widespread adoption, traditional IP networks are complex. Several researchers believed that current Internet architecture, including the OSI layer structure, had grown beyond what it was designed for. This was the primary motivation for the Stanford Clean Slate program, which was based on the premise that shortcomings in the Internet architecture were structural, and that incremental research would not cut it. One of the major outcomes of the program was SDN and OpenFlow. Given their common origin, SDN and OpenFlow are often used interchangeably.

While the Clean Slate Program is credited with doing much of the heavy lifting on SDN, the concept of separation of the control and data plane can be traced back to the Public Switched Telephone Network (PSTN). In 2004, the IETF proposed an interface standard called ForCES [286], which sought to decouple the control and forwarding functions of devices. Industry acceptance though was scant. With the Clean Slate Program providing new impetus, there was renewed interest in investigating avenues to address weaknesses in the Internet architecture.

Work on OpenFlow continued at Stanford in an academic setting. With advances being shown on research and production networks, industry partners such as NEC and HP started manufacturing OpenFlow enabled hardware. Parallel work conducted at Google added further impetus to SDN [159]. A milestone was reached when the Open Networking Foundation (ONF) was founded in 2011 to promote SDN and OpenFlow.

3.3.3 SDN Control Plane

The unified control plane of an SDN setup consists of one or more SDN controllers that use open APIs to exert control over the underlying vSwitches or forwarding devices. In addition to pushing forwarding rules to the vSwitches, the controllers also monitor the environment; thereby giving the controllers ability to have forwarding decisions integrated with real time traffic management. The controllers interact with the rest of the SDN infrastructure using three communication interfaces, commonly called the southbound, northbound and east/westbound interfaces. The separation in their functions is as follows:

- Southbound interface allows the controller to communicate, interact and manage the forwarding elements. While other proprietary solutions exist, OpenFlow is, by far, the most common implementation of the southbound interface. Amongst the proprietary solutions with non-trivial market share are onePK (Cisco) and Contrail (Juniper Networks). An alternate IETF standard, ForCES failed to gain much traction or adoption.

- Northbound interface enables applications in the application layer to program the controllers by making abstract data models and other functionalities available to them. The northbound interface can alternately be considered an API into the network devices. Unlike OpenFlow for southbound interfaces, there is no overwhelming market leader or accepted standard for northbound interfaces.

- East/Westbound interfaces are meant for communication between groups or federations of controllers. Similar to the northbound interfaces, there is yet to be a universally accepted standard.

3.3.4 SDN Data Plane

The data plane in the SDN architecture is tasked with enabling the transfer of data from the sender to the receiver(s). They are agnostic to the protocol that is used for communication between the end points. With the exception of communication with the controller, devices in the data plane themselves do not generate or receive any data, but instead act as conduits for data. Data plane devices need to support a southbound API, to communicate with the controllers. Devices in the data plane come in two flavors: 1) Software-based, such as Open vSwitch; and 2) Hardware-based such as a OpenFlow enables HP switch. As can be envisaged, software-based devices have a more complete feature set, but are generally slower.

3.3.5 OpenFlow

OpenFlow, defined by the ONF [8], is a protocol between the control and forwarding layers of an SDN architecture, and is by far the most widespread implementation of SDN. A basic OpenFlow architecture consists of end hosts, a controller and OpenFlow enabled switches. Note that contrary to the traditional network nomenclature, an OpenFlow *switch* is not limited to being a layer-2 device. The controller communicates with the switches using an OpenFlow API.

When a packet arrives at an OpenFlow switch, packets are processed as follows:

1. A flow table lookup attempting to match the header fields of the packet in question to the local flow table is done. If no matching entry is present, then the packet is sent to the controller for processing. When multiple entries that match the incoming packet are present in the flow table, the packet with the highest priority is

picked. Details about the match fields and actions are provided in Chapter 10.

2. Byte and packet counters are updated.

3. Action(s) corresponding to the matching flow rule is(are) appended to the action set. If a different flow table is part of the execution chain, then processing continues.

4. Once all flow tables have been processed, the action set is executed.

3.3.6 SDN Controllers

The controller is the brains of the SDN operation. It lies between the data plane devices on one end, and high level applications on the other. An SDN controller takes the responsibility of establishing every flow in the network by installing flow entries on switch devices.

Flow entries can be added to a data plane device in either a (1) proactive mode, where the flow rules are sent to the data plane devices as soon as the controller learns of it; or (2) reactive mode, where the controller sends flow entries to the data plane devices only as needed. In reactive mode, when data plane devices send flow setup requests to the controller, it first checks this flow against policies at the application layer and decides what actions need to be taken. Next, it determines a path for the packets to traverse (based on other application layer policies) and installs new flow entries in each device along the path. Flow entries that are added have specific timeout values which indicate to the data plane devices how long they should be stored in their forwarding tables in case of inactivity before removing the entry. The trade-off between setup delay and memory required to maintain the forwarding table in your environment determines the selection of the network administrator. Further, in reactive mode, the administrators are offered the ability to be agile according to the current network conditions.

In addition to the mode of operation, administrators face another design choice over flow granularity, where the trade-off lies between flexibility and scalability. This is akin to aggregated routes in traditional routers. While fine-grained flow rules offer flexibility and an additional layer of security, it can be unfeasible to implement. These aspects are discussed further in Chapter 10.

A few widely deployed OpenFlow-based SDN controllers are:

- NOX was among the first publicly available OpenFlow controller. Several variants of NOX exist today; such as NOX-MT, QNOX, Fort-Nox, POX, etc.; each with a different emphasis.

- OpenDaylight (ODL) is an open-source SDN controller that has been available since 2014. It is a modular multi-protocol SDN controller that is widely deployed in the industry. A more detailed exploration of ODL is shared in Section 3.3.9. Companies such as

ConteXtream, IBM, NEC, Cisco, Plexxi, and Ericsson are active contributors to ODL.

- OpenContrail is a flavor of SDN controller originally from Juniper Networks. It is an Apache 2.0-licensed project that works well with virtual routers, and has an analytics engine.

- Beacon is a JAVA-based OpenFlow control built at Stanford University. It has a multi-threaded, cross-platform, and modular controller.

- Floodlight is a JAVA-based OpenFlow controller originally forked from Beacon by Big Switch Networks. Similar to OpenContrail, it is also an Apache 2.0-licensed product that supports a range of virtual and physical OpenFlow switches. Further, it integrates well with OpenStack. It is supported and enhanced by companies such as Intel, Cisco, HP, and IBM.

- Ryu is a Python-based open-source SDN controller that is lightweight. It is modular in nature and supports several protocols such as OpenFlow, Netconf and OF-config.

- FlowVisor is a special-purpose OpenFlow controller that supports a decentralized controller. It allows for siloing of the resources to ensure only an assigned controller can control a switch. It also has traffic shaping functionalities.

In addition to installing flow rules, the controller is also used for traffic monitoring. In fact, traffic monitoring can be used in conjunction with application layer policies to implement rate limiting. The controller can obtain traffic from the data plane devices in a push-based or a pull-based fashion. While both accomplish the same result for the most part, each gives the administrator some positives over the other. For example, push-based mechanism is ill-suited for flow scheduling. Pull-based mechanisms need to be cognizant of the scalability and reliability of the centralized controller. Once again, the mode of choice is a design decision for the administrator.

3.3.7 Open Virtual Switch

OVS [9] is open-source implementation of a distributed programmable virtual multi-layer switch. OVS implementations generally consist of flow tables, with each flow entry having match conditions and associated actions. OVS communicates with the controller using a secure channel, and generally uses the OpenFlow protocol. OVS has been widely integrated into major cloud orchestration systems such as OpenStack, CloudStack [3] etc., in lieu of the traditional Linux bridge.

Figure 3.5 represents the main components of OVS. The kernel module receives the packets from a NIC (physical or virtual). If the kernel module knows how to handle the packet, it simply follows the instructions. If not, the packet is sent to the ovs-vswitchd in userspace using NetLink. This

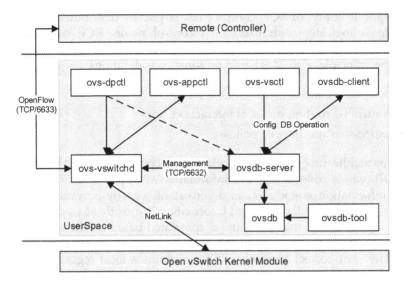

FIGURE 3.5
Open vSwitch architecture.

determines how the packet should be handled using the OpenFlow protocol. The ovs-vswitchd communicates with a ovsdb-server via a socket. The ovsdb-server stores OVS configuration and switch management data in JavaScript Object Notation (JSON) format. All functions in the userspace can be accomplished using Command Line Interface (CLI) commands.

OVS and vSwitch are used interchangeably in the remainder of this document.

3.3.8 Routing in SDN

In traditional networking, the topology information, i.e., connectivity and link weights between neighbors or Originator-Destination (OD) pairs, is propagated in the network for route computation. This usually results in high bandwidth, high process utilization, excessive storage occupancy, slow convergence, complex and unscalable networking. In an SDN environment, there is no device explicitly defined as a traditional router, where the control plane and data plane are separated. Link states are periodically polled by the controller to compute a configuration for each switch to direct packet switch in the data plane. As the result, routing in SDN is done in the SDN control plane. A few well-known routing strategies in SDN follow.

3.3.8.1 RCP: Routing Control Platform

Routing Control Platform (RCP) was proposed in [90] for the provisioning of inter-domain routing over a BGP network. In this architecture, routing is done as a separate entity. In RCP, control from physically distributed entities in a domain is logically centralized in a control plane. RCP computes and exchanges routes within the domain and with other domains' RCPs. The working principle of RCP is based on three considerations:

- network's consistent view of the state,
- controlled routing protocol interaction, and
- expressive and flexible policies.

In this architecture, legacy network routing constituent iBGP Route Reflector (RR), which collects network information swapped with RCP. The network information is gathered in a consistent way by accumulating eBGP learned routes. Then the collected information is directly shared with border routers. Afterwards, the best route is computed based on the eBGP learned route. The entire routing configuration and routing states reside in RCP control plane. For correct handling, RCP maintains a local registry which has global view and information is exchanged with other RCPs for inter-domain routing. The implementation of RCP in SDN is known as Intelligent Route Service Control point (IRSCP) [273]. IRSCP is an architecture that was in the use before the emergence of SDN.

3.3.8.2 The SoftRouter

The SoftRouter [174] architecture is presented with the aim of separation of control and forwarding elements called Control Element (CE) and Forwarding Element (FE), respectively. The control functionality is provided by using a centralized server, i.e., a CE that might be many hops away from the FE. The SoftRouter architecture is inspired by softswitch [279] concept of telecommunication industry, where software-based switching is done for separating the voice from the control path. Many FEs may be connected to one CE. The communication between CE and FE is done through ForCEs [82]. The control protocol runs over the CE but the topology discovery module is run by the FE to form an NE. When SoftRouter is employed in a BGP-based AS, IGP route computation is done in CE using ForCES protocol. In contrast to RCP, where intra-AS route computation is done using IGP. SoftRouter also uses iBGP for inter-mesh AS connectivity.

3.3.8.3 RF IP Routing: IP Routing Services over RouteFlow-based SDN

RouteFlow was a project initially named as QuagFlow [200], which aimed to provide IP routing as Router-as-a-Service in a virtualized environment. The RouteFlow basically consists of three components: RouteFlow Controller, RouteFlow Server, and Virtual Network Environment.

RouteFlow controller interacts with RouteFlow Server through RouteFlow Protocol. The Virtual Environment (VE) consists of RouteFlow Client and

Routing Engine. The RouteFlow Client collects Forwarding Information Base (FIB) from the Routing Engine (Quagga, BIRD, XORP, etc). RouteFlow Client transforms these FIBs into OpenFlow tuples that are forwarded to the Route-Flow Server that is responsible for establishing routing logic out of these tuples. The routing logic is transferred to RouteFlow controller that defines the match field and action against each match. The VE is connected directly to the Route-Flow controller through a Virtual Switch, such as OVS. The direct connection between VE and controller reduces the delay by providing a direct mapping of physical and virtual topology. There were no databases used in the first phase of the development of RouteFlow, which may choke the RouteFlow Server. To overcome this issue, NoSQL (MongoDB, Redis, CouchDB) was introduced in the RouteFlow architecture to provide inter-process communication between different components of the architecture. RouteFlow performs multiple operations in different scenarios: i) logical split, ii) multiplexing and iii) aggregation. All routing tasks are done by the virtual environment that provides flexibility. The different phases of RouteFlow development makes it possible to integrate it with SDN, so much so that RouteFlow is considered the basic architecture to control routing in SDNs.

3.3.8.4 VRS: Virtual Routers as a Service

In [52], the authors describe an architecture characterized by a virtual routing instance that is responsible for managing distributed forwarding elements termed as Virtual Router System (VRS). Virtual router instances communicate with a Point-Of-Presence (POP) and follow a star topology, in which a single core node is connected to Customer Edge Gateways (CEG) linked through Intermediate Nodes (INs). VRS instances are associated with Forwarding Engines, which can be programmed using OpenFlow and flow tables are updated accordingly. The routing decision is computed in the core node. A VM instance associated with virtual routing controller in core install all the rules in the forwarding plane. Another module of this architecture is the Path Management Controller (PMC) which is used to calculate the minimum cost route. PMCs also manage data path migration. VRS generally involves two operations:

- core node location selection, and
- optimal forwarding path allocation.

VRS adheres to the following considerations while calculating minimum cost path:

- customer's geographical attachment,
- bandwidth demand, and
- corresponding capacity.

By keeping a view of these considerations, a node with maximum capacity and minimum cost data path is selected. The cost of VRS increases with the increase in CEG nodes.

3.3.8.5 RFCP: RouteFlow Routing Control Platform over SDN

In [233], the authors proposed RouteFlow Control Platform (RFCP), which is a hybrid networking model of two former studies, i.e., Routing Control Platform and RouteFlow. RFCP is an additional computational layer for computing routes within and between ASes. The RFCP adds a flavor of data store-centric platform that stores: i) RFCP Core state information, ii) network View, and iii) Network Information Base (NIB). The communication between NOS running SDN controller (POX, NOX, Maestro, Trema). Interfacing with the Virtual Environment is done through OpenFlow protocol control message. RFCP comprises of RouteFlow-Client, RouteFlow-Server, and Route-Flow-Proxy. To integrate this architecture with a traditional network, Route Reflector (RR) in BGP domain is interconnected with Provider Edge (PE) routers via iBGP, which communicates with BGP controller that is also called RFCP-Controller. The RouteFlow-controller serves as a gateway between Route Reflector and OpenFlow switches. RouteFlow Client gathers routing information from the routing engine that has a virtual image of physical topology. On the basis of collected information, routing logic is established in the RouteFlow-Server. RF-Proxy (RF Controller) exchanges network state between the switch and RFCP. The direct communication between the controller and virtual machines is done through OpenFlow control messages. Automatic configuration of RFCP has been made possible by the introduction of discovery controller. The enhanced version now consists of five modules: i) RouteFlow controller, ii) topology controller, iii) Remote Procedure Call (RPC) client, iv) RPC server, and v) FlowVisor. Topology controller monitors the entrance of new switch, i.e., change in topology, executes the topology discovery module and directs this configuration information to RPC client. The configuration contains switch ID and port number. On the basis of this configuration information, RPC server generates a VM of the same ID and with same port resembling the physical topology network. This VM is assigned with IP addresses allocated by the topology controller. All information is stored in the database by the RF-controller in the form of configuration files. These configuration files are passed to the RPC server through RPC Clients that configure this information in created virtual machines.

3.3.8.6 RaaS: Routing as a Service

In [151], authors proposed a Routing Service as an intelligent application based on OpenFlow architecture. A logically centralized routing control plane is used, which has a complete view of the network topology. This global view enables the centralized control to make a routing decision. The building blocks of a centralized routing plane are

- Link Discovery Module: determines the physical links between OpenFlow enabled devices;
- Topology Manager: upholds network status information, i.e., topology; and
- Virtual Routing Engine: creates a virtual topology for incorporating traditional routing over a virtual routing engine.

The routing controller initiates routing services by advertising its information and sending packet-out messages on all the connected OpenFlow switches' ports. In this way, the flow look-up procedure is done across all connected OpenFlow switches and flow entries are populated using packet-in message sent towards the centralized controller. For a non-OpenFlow switch, a packet is broadcast as in traditional link discovery protocol. Routing decisions are based on the information stored in the database attached to the centralized routing plane and routing decision accuracy is determined by keeping track of MAC addresses and port numbers of each connected device. The best route is calculated using the Dijkstra algorithm.

3.3.8.7 CAR-Cloud Assisted Routing

An architecture that provides RaaS is presented in [144]. The proposed architecture is Cloud Assisted Routing (CAR), in which routing complexity is handled by using clouds based on two principles: i) CPU Principle: Keep control plane closer to the cloud, i.e., transferring computational intensive routing function on cloud; and ii) Memory Principle: keeping data plane closer to the cloud so as to place rarely used prefixes at the cloud. The architecture of the CAR includes two types of routers: i) hardware routers: keeping partial FIB, and ii) software routers: keeping full FIB. The authors basically present an architectural framework for shifting CPU intensive computation into the cloud to leverage cloud computation and memory benefits. How much routing computation is shifted on the cloud is a basic consideration in this architecture. The inter-domain routing in a BGP-based AS can easily be shifted on this architecture by shifting Route Reflector into the cloud.

3.3.9 OpenDaylight

OpenDaylight (ODL) is an open-source project under the Linux Foundation [7]. Applications running on the ODL controller use a Service Abstraction Layer (SAL) to communicate with different types of devices using a variety of communication protocols, and provide RESTful APIs for use by external applications. ODL was chosen as the controller in this implementation

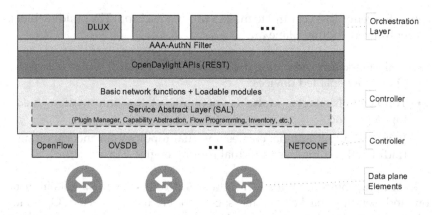

FIGURE 3.6
ODL architecture.

because of its large open-source development community, as well as indications during decision making that ODL would be adopted as an industry standard. This work extends the stable Lithium version of the controller. Figure 3.6 shows the ODL architecture including the different modules.

The ODL project repository, available at [16], follows a microservices architecture to control applications, protocols, plugins and interfaces between providers and customers. It uses YANG data structures along with shared data stores and messaging infrastructure to implement a Model Driven SAL (MD-SAL) approach to solving more complex problems. This model helps keep the controller as lightweight as possible, providing users with the ability to install protocols and services as needed. As of this book, the ODL ecosystem has implementations for Switching, Routing, Authentication, Authorization and Accounting (AAA), a DLUX-based Graphical User Interface (GUI) and support for protocols such as OpenFlow, NETCONF, BGP/PCEP, SNMP, CAPWAP. Additionally, it interfaces with OpenStack [207] and OVS through the OVSDB Integration Project [214]. This modularization and separation of functionality has been implemented per the Open Services Gateway Initiative (OSGi) specification, and as such provides for service object initiation, dynamic module handling and graceful exit.

ODL uses Apache Karaf [2] as its OSGi container. Applications[2] in Karaf are independent of each other, and can be started, stopped or restarted without affecting other applications. Brew uses the `l2switch`, `openflowplugin`, `openflowjava`, `yangtools`, `netconf` and `dlux` features. RESTCONF [47] provides a RESTful API to perform Create, Retrieve, Update and Delete (CRUD) operations using NETCONF, which itself is a means to configure network elements in a vendor-agnostic manner using the YANG modeling

---

[2] Interchangeably called bundles or features. Karaf command line uses the keyword *feature*.

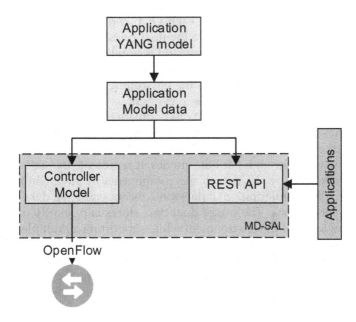

FIGURE 3.7
MD-SAL application development.

language. Figure 3.7 shows the relationship between the different protocols and modeling languages in a MD-SAL development paradigm [201]. A new application development requires defining the application's model using YANG.

ODL maintains two different data stores, as shown in Figure 3.8. Classified broadly on the type of data maintained in them, they are: *a*) configuration data store; and *b*) operational data store. Since the data is stored in a tree format, the configuration and operational data stores are interchangeably called the configuration and operational trees.

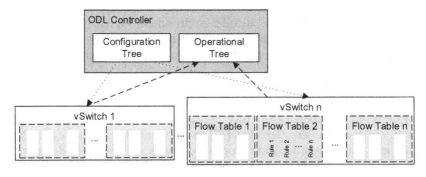

FIGURE 3.8
ODL data stores.

The configuration data store on each ODL controller contains data that describes the changes to be made to the flow rules on the switches. It represents the intended state of the system, and is populated by administrators or applications on the controller. The configuration data store contains information about every device present in the environment, flow tables associated with the devices, and the flow rules in every flow table. To give administrators and other applications the ability to populate this data store, it has read/write permission. The operational data store matches the configuration data store in structure, but contains information that the controller discovers about the network through periodic queries. It represents the state of the system as understood by the data plane components in the environment. As opposed to the configuration data store, the operational data store has read-only permissions. The use of dual data stores is primarily to maintain global knowledge of the environment while supporting a multiple controller scenario. For example, if Controller 1 has a new flow rule that is used by an OVS to direct traffic, Controller 2 would learn of this flow rule when it populates its operational data store with all the flow rules present in the environment. This would happen irrespective of the communication between the two controllers.

3.3.10 Distributed SDN Environments

Distributed controller environments in SDN are widely studied. Onix [159] facilitates distributed control in SDN by providing each instance of the distributed controller access to holistic network state information through an API. HyperFlow [267] synchronizes the network state among the distributed controller instances while making them believe that they have control over the entire network. Kandoo [289] is a framework tailored for a hierarchical controller setup. It separates out local applications that can operate using the local state of a switch, and lets the root controller handle applications that require network-wide state. DISCO [216] is a distributed control plane that relies on a per domain organization, and contains an east-west interface that manages communication with other DISCO controllers. It is highly suitable for a hierarchically decentralized SDN controller environment. ONOS [45] is an OS that runs on multiple servers, each of which acts as the exclusive controller for a subset of switches and is responsible for propagating state changes between the switches it controls.

Dixit et al. [81] presented an approach to dynamically assign switches to the controllers in a multiple controller environment in real-time. The balanced placement of controllers can reduce the cost and the overhead for dynamic assignment of controllers. Bari et al. [37] also presented a technique to dynamically place controllers depending on the changes of number of flows in the network. Controller placement problems have been studied extensively from a performance perspective [37, 156, 288, 290], and based on resilience [40, 123, 124, 272, 282].

3.3.11 Distributed SDN Controller Considerations

SDN was designed with a centralized control plane in mind. This empowers the controller with a complete network-wide view and allows for the development of control applications and for easier policy enforcement. Centralizing the control plane in SDN is fraught with scalability challenges associated with the SDN controller being a bottleneck [290]. Benchmarking tests on an SDN have shown rapid increase in the performance of a single controller, from about 30,000 responses per second using NOX [264] in 2009 to over 1,350,000 responses per second for Beacon [87] in 2013. But with data center architectures dealing with 100 GB network traffic (equal to about 130 Million Packets Per Second (MPPS) [12]), a single controller would still not scale well enough to be deployed in a cloud environment [42]. Further, large production environments still demand performance and availability [45]. Distributing the controller responsibilities to multiple devices/applications, while maintaining logical centralization is an obvious solution. Figure 3.9 shows a representation of major different distributed controller categories; namely clustered and hierarchical.

While the OpenFlow protocol supports multiple controller environments, the controllers themselves need to be able to: *a*) allow a switch to establish communication with them; and *b*) have mechanisms in place to process handover, fail overs, etc. OpenFlow shields itself from the complexities of a multiple controller environment, and just requires the controller to have one of three roles - OFPCR_ROLE_MASTER, OFPCR_ROLE_EQUAL or OFPCR_ROLE_SLAVE[3].

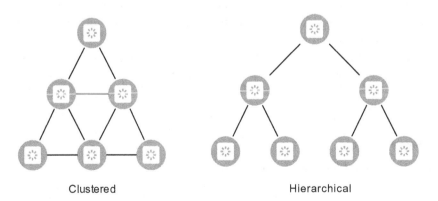

Clustered   Hierarchical

FIGURE 3.9
Distributed controller classes.

---

[3] The *Master* and *Slave* roles are self-explanatory. *Equal* role and *Master* role are exactly the same, with the difference that only one controller can be *Master* at a time, while multiple controllers can share the *Equal* role.

Moving to a distributed controller environment essentially splits the roles of the SDN controller between multiple devices that communicate with the switches, and a data store that retains complete environment knowledge.

3.3.12 Challenges in Multiple-Controller Domain

Several studies have attempted to study distributed SDN controllers. However, despite their attempt at distributing the control plane, they require a *consistent* network wide view in all the controllers. Maintaining synchronization and concurrency in a highly dynamic cloud environment is problematic. Since the SDN switches look to the controllers for answers while handling unfamiliar flows, knowing which controller to ask is important. Moreover, the controllers themselves need to have methodologies to decide who controls which switch, and who reacts to which events. And most of all, consistency in the security policies present on the controller is paramount - the absence of which might result in attackers using application hopping across multiple partitions of the SDN environment without permissions.

Since one of the primary motivations behind SDN was a centralized control plane that has complete knowledge of the environment, maintaining a complete picture after dividing the network into multiple subnetworks requires information aggregation. This could be challenging, especially when the environments are dynamic.

While designing a distributed controller architecture, the implications of controller placement need to be considered carefully. For example, security policies in a mesh controller architecture would have to ensure minimal address space overlap; while in a hierarchical architecture, it may be acceptable for the lower level (leaf) controllers to share address spaces. In environments where latency is a concern, the distance between controllers and the switches needs to be minimized.

Finally, studies also suggest that distributed control planes are not adaptable to heterogeneous and constrained network deployments [216]. This removes a certain amount of desired design flexibility from the SDN setup.

3.4 Advanced Topic: Deep Programmability

The recent developments in the field of SDN saw emergence of protocols similar to OpenFlow that could widen the movement towards adoption of SDN. Programming Protocol Packet Processors (P4) [50] is one such language. The OpenFlow protocol originally started with Layer 2 headers that allowed processing of packets traversing through OpenFlow capable switches. The number of headers have grown from 12 to 41 in the latest version of OpenFlow. This has increased the expressive capability of the protocol, but at the same time increased the complexity of adding new headers. Different cloud and data-center service providers need additional headers to the current Open-Flow protocol, in order to support encapsulation (e.g., VXLAN, NVGRE, and STT). The P4 proposes a flexible mechanism for the parsing of packets and matching header fields. The goals of P4 are

1. Reconfigurability, allowing programmers to change the way in which switches process the packets, once they are deployed.
2. Protocol independence, where switches should not be tied to specific network protocols. The packet parser in the OpenFlow switches is fixed, whereas P4 allows creation of custom parser for different types of network traffic. Additionally, P4 consists of a collection of typed match+action tables to process the packets.
3. Target independence, wherein the packet processing capability should be independent of the underlying hardware.

3.4.1 P4 Forwarding Model

P4 digresses from the fixed switch design in OpenFlow protocol. The recent chip designs allow low-level chip interface to be exposed in form of abstract interface for the programmers to configure the switch behavior at terabit speeds. P4 acts as a middleware between the switches and SDN controller, allowing the SDN controller to change the way switch operates. P4 forwarding model consists of two main operations, i.e., *Configure* and *Populate*. The configuration operation creates a parser for the network traffic, sets the order of match+action stages, and specifies the header fields to be processed at each stage. It also specifies the protocols supported and other packet processing operations. The populate operation adds/removes the entries from match+action tables created during the configuration phase.

As shown in Figure 3.10, the packets arriving are handled first by the parser. The header fields of the parser are extracted and the protocols supported by the switch are identified. The packet body is buffered separately. The extracted header fields are then passed to the match+action tables to determine the packet egress port and the queue that is processing the packet. Based on the match+action, the packet may be forwarded, dropped, rate limited or replicated. The packet is then passed to egress match+action for per-instance flow modification, frame-to-frame flow state tracking. Packet can also carry additional metadata for information such as packet scheduling, virtual network identifier, etc.

3.4.2 P4 Programming Language

P4 follows control flow programming paradigm to describe header field processing using declared header types and a primitive set of actions. P4 language

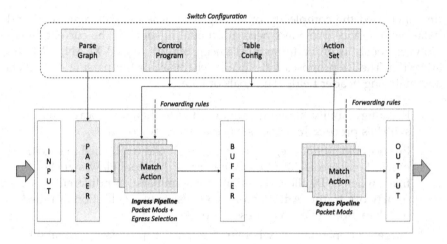

FIGURE 3.10
P4 abstract forwarding model.

allows identification of sequential dependencies between header fields of the packet as shown in the Figure 3.11, and optimize the packet processing by executing some operations in parallel. The Table Dependency Graphs (TDGs) shown above help in describing field inputs, actions and control flow between tables.

P4 consists of a two-step compilation process. At higher level the programmers express the packet processing using imperative control flow language (P4), the control flow is then translated into TDG, which facilitates dependency analysis and mapping of TDG to a specific target switch. The detailed

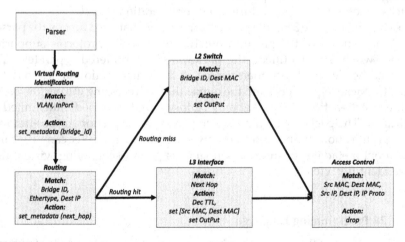

FIGURE 3.11
P4 table dependency graph (TDG) for L2/L3 switch.

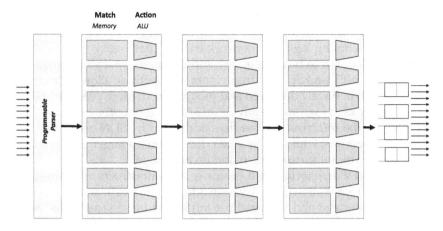

FIGURE 3.12
PISA architecture.

programming repository and tutorials for testing P4 language can be analyzed by the interested reader by cloning P4 git repository [51].

3.4.3 Protocol Independent Switch Architecture

The network switches in modern data centers have fixed functions. The flexibility required by network programmers in order to modify the behavior of the network switches comes at a cost of performance. Protocol Independent Switch Architecture (PISA) is a new paradigm for processing packets at high speed, under the programmatic control of the user.

As shown in the Figure 3.12, PISA [76, 253] consists of about 11 primitive instructions that can be programmed using very uniform pipeline of programming stages to process packet headers in rapid succession. The programs are written in P4 language, and compiled by Barefoot Capilano compiler, which allows line-rate packet processing on the PISA device. Example of PISA device includes Barefoot Network's Tofino switch which operates at 6.5Tb/s, and is fully programmable via P4 programming language.

Summary

The need for dynamism in allocating resources to changing and growing need for networking capability to support compute capability gave rise to two complementary paradigms - NFV and SDN. They both allow the network administrators to manage network services through abstraction of lower-level functionalities, as well as provide programmable APIs allowing computer networking applications to interact with networking functions based on predefined programming logic and network traffic situations.

While both sought to bring in agility and dynamism, they are fundamentally different. NFV and SDN address common objectives: allow network services to be automatically deployed and programmed. More precisely, SDN is a tool typically used for dynamically establishing a connection between VNFs. Furthermore, the infrastructure services addressed by SDN are rather basic connectivity services, while the NFVI services address a larger scope and provide a framework for virtualization and orchestration.

Thanks to NFV architecture and processes, the infrastructure services layer is decoupled from the resources layer. In addition, network and IT resources will soon be allocated and managed per infrastructure services. Thus, by isolating IoT use-cases, the risks to impact other infrastructure services, flood or break down the whole infrastructure is prevented.

4

Network Security Preliminaries

Network security consists of the policies and practices adopted to prevent and monitor unauthorized access, misuse, modification, or denial of a computer network and network-accessible resources. Network security involves the authorization of access to data in a network, which is traditionally controlled by the network administrator. To achieve the goal of network security, one must first understand attackers, what could become their targets, how these targets might be attacked, what gains attackers can derive, and consequences that the victim will suffer. Thus, how to model attacks is the first target of this chapter. The tasks of network security are to provide confidentiality, integrity, nonrepudiation, and availability of useful data that are transmitted in public networks or stored in networked computers. Building a deep layered defense system is the best possible defense tactic in network security. Within this type of defense system, multiple layers of defense mechanisms are used to resist possible attacks. Follow this strategy, we also presented layered security approaches by describing network reconnaissance, preventive defense solutions, and network security monitoring and management foundations. By finishing this chapter, readers should have a fundamental understanding of computer network security.

The rest of this chapter is arranged as follows: In Section 4.1, basic concepts of computer network security are presented including threat and attack models, layered security defense, and cyber killer chain; network mapping and port scanning based network reconnaissance are presented in Section 4.2; preventive network security approaches such as firewall and IPS are described in Section 4.3; network intrusion prevention and secure logging services are presented in Section 4.4; and finally, the network security assessment approaches are presented in Section 4.5.

4.1 Basic Concepts of Computer Network Security

4.1.1 Threat, Risk, and Attack

In computer security, a threat is a possible danger that might exploit a vulnerability to breach security and therefore cause possible harm. A threat can be either "intentional" (i.e., hacking from an individual cracker or a criminal organization) or "accidental", e.g., the possibility of a computer malfunctioning.

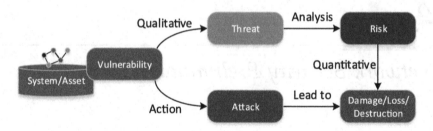

FIGURE 4.1
Security terms: vulnerability, threat, attack, and risk.

In the area of computer network security, threat modeling is about risk management in their core, and thus, they are often used in the same conversations. The terms threat and attack are often used interchangeably, which most often leads to incorrect interpretation of their meanings. Generally speaking, a threat is often due to the existence of vulnerabilities in a system that can be explored through actual attack actions. Thus, it is important to understand the relation among terms: threat, risk, vulnerability, and attack. Their relation is highlighted in Figure 4.1. In short, the relation among vulnerability, threat, attack, and risk can be explained as follows:

- A *vulnerability* is a weakness or gap in a security system that can be either exploited by attackers or caused by malfunctioning system components.
- A *threat* is the possibility of exploration of vulnerabilities that can lead to something bad happening, and it emphasizes the *qualitative* of potential damages due to explored vulnerabilities.
- An *attack* is an action triggered by deploying an attacking method, when a vulnerability is exploited to actually realize a *threat*.
- A *risk* is the quantifiable likelihood of loss due to a realized threat, and it emphasizes the *quantitative* of potential damages.

As shown in Figure 4.1, the two terms that get mixed up most often are threat and attack. They both aim to achieve damages, losses, or destruction of a given computer system. However, they focus on different security modeling aspects. Generally speaking, when we try to model vulnerabilities from an adversarial viewpoint, we are doing attack modeling but not threat modeling. For example, when we start with a vulnerability, and see what kind of damage it can lead to through deploying attacks, we basically model an attack. This is how traditional "bug hunting" operates.

For threat modeling, it is generally viewed from the defender's viewpoint, in which we start with a vulnerability, and see how likely the vulnerability can be explored, i.e., qualitative analysis. In formal terms, threat modeling is the process of identifying a system (or assets), potential threats against the system. It is thinking ahead of time about what could go wrong and acting accordingly. Through a quantitative measurement model if we can build one, we can usually derivate a risk analysis model for a given explorable vulnerability. As a result, risk analysis is usually performed offline before a damage can happen based on a given system/asset containing one or multiple vulnerabilities.

When evaluating a countermeasure of an attack, we usually use attack model to describe four interdependent terms:

- *The source of attacks*: there are two types of attackers based on their origins called inside attacker (or insider) and outside attacker (or outsider). For inside attackers, they are usually initiated within the security perimeter by an authorized user, e.g., a privileged DBA copying customer information and selling it. For outside attackers, they are usually initiated from the outside of security perimeter by an unauthorized user, e.g., an attacker performing a SQL Injection attack via a vulnerable application.

- *The method of attacks*: there are two types of attack methods based on the intention of the attacker called passive attacks and active attacks. For passive attacks, attackers do not alter resources while trying to learn information, e.g., wiretapping, port scanning, etc. For active attacks, attackers alter resources, e.g., IP spoofing, DDoS attacks, buffer overflows attacks, etc.

- *The target of attacks*: it is the objective of attackers to try to compromise. For example, an attack target can be a program, a computer system, a network segment, etc.; in other words, it is also called victim in computer security.

- *The consequence of attacks*: it describe outcomes by successfully deploying an attack. It is a multi-faceted consequence. From the functional aspect, the attacking target can be malfunctioned, e.g., not working or wrong function logics. From the social aspect, the victim (or the victim owner) will lose reputation as a result of security breaches. From the economic aspect, the cost for the security breach remedy can be prohibitively high.

4.1.2 Defense In Depth

Defense in depth (also known as Castle Approach [197]) is the concept of protecting a computer network with a series of defensive mechanisms such that if one mechanism fails, another will already be in place to thwart an attack [184]. Because there are so many potential attackers with such a wide variety of attack methods available, there is no single method for successfully protecting a computer network. Utilizing the strategy of defense in depth will reduce the risk of having a successful and likely very costly attack on a network.

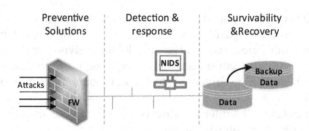

FIGURE 4.2
Layers of computer network security.

Generally speaking, the layers of computer network security can be broadly presented as three-layer defense in depth framework, which is presented in Figure 4.2. The first layer is a prevention mechanism that stops attacks from getting into the networking system. Inevitably, some attacks can defeat the prevention mechanisms. That is, there are "holes", or, weaknesses, in the prevention mechanism that allows some attacks to get through. For example, a packet filtering Firewall (FW) cannot prevent application-level attacks, e.g., downloading a malicious code that can be triggered in the future.

The second layer is detection and response mechanisms that watch activities on systems and networks to detect attacks and repair the damages. Again, there will be attacks that can go undetected, at least for a while. For example, attacks that blend in with normal activities, such as the Advanced Persistent Threat (APT) malware can be a malicious browser plug-in, in which it is hard to detect initially, not until its effects, such as data loss due to stolen credentials, manifest sometime later.

The third layer is attack-resilient technologies that enable the core elements, or, the most valuable systems, on the network to survive attacks and continue to function. For example, a database server is a collection of diversified systems with redundant backups, so that compromised data can be recovered after the attack created damages in one database system.

4.1.3 Cyber Killer Chain

The term "kill chain" is a term used originally by the military to define the steps the enemy uses to attack a target. In 2011 Lockheed Martin released a paper defining a "Cyber Kill Chain" [180] that adopts the concept of a procedural step-by-step attacking method consisting of target identification, force dispatch to target, decision and order to attack the target, and finally the destruction of the target. Particularly, the cyber kill chain includes the following steps:

- *Reconnaissance*: The attacker gathers information on the target before the actual attack starts. Attackers collect information on their intended targets by searching the Internet, social network sites such as LinkedIn or Facebook. In addition, they may try to gather information through social engineering approaches such as employee calling, email interactions, dumpster diving, etc. Operations Security (OPSEC) [278] plays a big defending role here: trained and aware workforces will know they are a target and limit what they publicly share; they will authenticate people on the phone before they share any sensitive information; they safely dispose of and shred sensitive documents; etc. At the network level, networking mapping and port scanning are fundamental approaches that an attack can understand the targeting network's topology and running services. They can be followed by vulnerability exploration by using various penetration tools.

- *Weaponization*: Cyber attackers do not interact with the intended victim, instead they create their attack. For example, the attacker may create a malicious code, e.g., network worm, for later sending to an identified vulnerable network port in the targeting system.

- *Delivery*: Transmission of the attack to the intended victim(s). For example, this would be sending the actual malicious codes to a victim network node.

- *Exploitation*: This implies actual detonation of the attack, such as the exploit running on the system. For example, the explored vulnerable system may suffer buffer-overflow problem and the attacker may gain root privilege of the exploited system. Then, the attacker can do further malicious actions on it, such as erase the trace, install back-door software, etc.

- *Installation*: The attacker may install malware on the victim. However, not all attacks require malware, such as a CEO Fraud Attack, harvesting login credentials, or stealing secret data.

- *Command & Control (C&C)*: This implies that once a system is compromised and/or infected, the system has to call home to a C&C, a Command and Control system for the cyber attacker to gain control, which is usually done through the installed back-door program.

- *Actions on Objectives*: Once the cyber attackers establish access to the organization, they then execute actions to achieve their objectives/ goal. Motivations greatly vary depending on the attacking goal, to include political, financial or military gain. Nowadays, attackers are quite commonly using the compromised system as a stepping stone to explore the next victim. In such, the traditional strong north-south bound traffic control, in which the security is heavily depending on perimeter firewalls will not work well. Attackers may laterally explore vulnerable systems through east-west traffic within the secure perimeter.

Cyber killer chain is a comprehensive approach to model attackers' behavior based on a step-by-step procedural approach. However, it has limitations in that attackers may not follow the identified step, and sometimes, they may change the sequence [160]. For insider attacks, some steps can be skipped; and for Advanced Persistent Threat (APT) attackers, they can explore the system through a long duration, which cannot be easily identified and prevented using traditional network security solutions. However, understanding attackers' potential actions and goals is helpful to construct security solutions targeted at each step.

4.2 Network Reconnaissance

4.2.1 Network Mapping

Network mapping is the study of the connectivity of networks at the layer 3 on a TCP/IP network. Network mapping discovers the devices, i.e., virtual or physical, on the network and their connectivity. There are two prominent techniques used today for network mapping. The first works on the data plane of the network and is called active probing. It is used to infer network topology based on router adjacencies. The second works on the control plane and infers autonomous system connectivity based on network management system, e.g., SNMP (Simple Network Management Protocol).

Active probing relies on *traceroute*-like probing on the IP address space. These probes report back IP forwarding paths to the destination address. By combining these paths, an attacker can infer router level topology for a given scanning network. Active probing is advantageous in that the paths returned by probes constitute the actual forwarding path that data takes through networks. However, active probing requires massive amounts of probes to map the entire network and it may miss some links/paths that are not on the shortest path of probes. It is also more likely to infer false topologies due to load balancing routers and routers with multiple IP address aliases. Decreased global support for enhanced probing mechanisms such as source-route probing, ICMP Echo Broadcasting, and IP Address Resolution techniques leaves this type of probing used mainly in the realm of network diagnosis.

Using network management framework such as SNMP is another way to achieve the network mapping. However, SNMP v3 [258] has strong security protection, which is difficult for external attackers to compromise the management system. Other auditing systems such as network logs can be potentially attacking targets.

Through network mapping, attackers aim to find out whichever systems they can reach, and their connectivity to other victims that cannot be directly reached from attackers. If the attackers have no access to your internal network, they will begin by mapping and scanning your Internet gateway, including your DMZ systems, such as Internet-accessible Web, Mail, FTP, and DNS servers. They will methodically probe these systems to gain an understanding of your Internet perimeter. After conquering your perimeter, the attackers will attempt to move on to your internal network.

4.2.2 Port Scanning

After attackers know the addresses of live systems on the network and have basic understanding of your network topology, next they want to discover the purpose of each system and learn potential entryways into the victim hosts by analyzing which ports are open. Active TCP and UDP ports on a host are indicative of the services running on those systems. Each machine with a TCP/IP stack has 65,535 TCP ports and 65,535 UDP ports. Every port with a listening service is a potential doorway into the machine for the attacker, who will carefully take an inventory of the open ports using a port-scanning tool.

For example, a web server listens to TCP port 80 for unencrypted traffic and TCP port 443 for encrypted traffic that are default ports assigned to the web service. When running a DNS server, UDP port 53 will be open. If the machine is hosting an Internet mail server, TCP port 25 is likely open. RFC1700, Assigned Numbers, which contains a list of these commonly used port numbers and now assigned port numbers are maintained by an online database by IANA [69].

4.2.3 Vulnerability Scanning and Penetration Testing

Vulnerability scanning is an inspection of the potential points of exploit on a computer or network to identify security holes. For security enhancement purposes, a vulnerability scan detects and classifies system weaknesses in computers, networks and communications equipment and predicts the effectiveness of countermeasures. A scan may be performed by an organization's IT department or a security service provided possibly as a condition imposed by some authority. From the attacking aspect, vulnerability scans are also used by attackers looking for vulnerabilities as points of entry to hack into the system.

A vulnerability scanner (such as Nessus, Open VAS, GFI LANGuard, Rapid7, Retina, Qualys) usually runs from the end point of the person inspecting the attack surface in question. The software compares details about the target attack surface to a database of information about known security holes in services and ports, anomalies in packet construction, and potential paths to exploitable programs or scripts. The scanner software attempts to exploit each vulnerability that is discovered. Running a vulnerability scan can pose its own risks, as it is inherently intrusive on the target machines running code. As a result, the scan can cause issues such as errors and reboots, reducing productivity.

Generally speaking, there are two approaches to perform a vulnerability scan, namely, unauthenticated and authenticated scans. In the unauthenticated method, the scanner performs the scan as an intruder would, without trusted access to the network. Such a scan reveals vulnerabilities that can be accessed without logging into the network. For an authenticated scan, the scanner logs in as a network user, revealing the vulnerabilities that are accessible to a trusted user, or an intruder that has gained access as a trusted user.

Another quite often used security term is penetration testing, which can be easily confused when distinguishing it from vulnerability scanning. Indeed, penetration testing is quite different, as it attempts to identify insecure business processes, insecure system settings, or other weaknesses. For example, transmission of unencrypted passwords, password reuse, and forgotten databases storing valid user credentials are examples of issues that can be discovered by a penetration test. Penetration tests do not need to be conducted as often as vulnerability scans but should be repeated on a regular basis.

Penetration tests are best conducted by a third-party vendor rather than internal staff to provide an objective view of the network environment and avoid conflicts of interest. Various tools are used in a penetration test, but the effectiveness of this type of test relies on the tester. The tester should have a breadth and depth of experience in information technology, preferably in the organization's area of business; an ability to think abstractly and attempt to anticipate threat actor behaviors; the focus to be thorough and comprehensive; and a willingness to show how and why an organization's environment could be compromised.

A penetration test report should be short and to the point. It can have appendices listing specific details, but the main body of the report should focus on what data was compromised and how. To be useful for the customer, the report should describe the actual method of attack and exploit, the value of the exploited data, and recommendations for improving the organization's security posture. In Table 4.1, a comparison is given to describe the difference between vulnerability scan and penetration test.

4.3 Preventive Techniques

4.3.1 Firewalls

In computer security, a firewall (FW) is a component or set of components that restricts access between a protected network and the Internet, or between other sets of networks. In Chapter 6, we provide a more advanced topics of firewalls. In this section, we focus on technical background of firewalls. Some frequently used terms that are highly related to firewalls are given below:

TABLE 4.1

Comparison of Vulnerability Scans Versus Penetration Tests

	Vulnerability Scan	**Penetration Test**
Frequency	More frequently and can be ad hoc, especially after new equipment is loaded or the network undergoes significant changes	Once or twice a year, as well as anytime the Internet-facing equipment undergoes significant changes
Reports	Provide a comprehensive baseline of what vulnerabilities exist and what changed since the last report	Concisely identify what data was compromised
Focus	Lists known software vulnerabilities that could be exploited	Discovers unknown and exploitable weaknesses in normal business processes
Performed by	Typically conducted by in-house staff using authenticated credentials; does not require a high skill level	Best to use an independent, outside service and alternate between two or three; requires a great deal of skill
Value	Detects when equipment could be compromised	Identifies and reduces weaknesses

- *Host*: A computer system attached to a network.

- *Dual-homed host*: A dual-homed host is a term used to reference a type of firewall that uses two (or more) network interfaces. One connection is to an internal network and the second connection is to the Internet. A dual-homed host usually runs general purpose operating system to support firewall applications running in it.

- *Network Address Translation (NAT)*: A procedure by which a router changes data in packets to modify the network addresses. This allows a router to conceal the addresses of network hosts on one side of it. This technique can enable a large number of hosts to connect to the Internet using a small number of allocated addresses or can allow a network that is configured with illegal or un-routable addresses to connect to the Internet using valid addresses. It is not actually a security technique, although it can provide a small amount of additional security. However, it generally runs on the same routers that make up part of the firewall.

- *Perimeter network*: A network added between a protected network and an external network in order to provide an additional layer of security. A perimeter network is sometimes called a DMZ, which stands for De-Militarized Zone (named after the zone separating North and South Korea).

- *Proxy*: A program that deals with external servers on behalf of internal clients. Proxy clients talk to proxy servers, which relay approved client requests on to real servers, and relay answers back to clients.

A packet filtering firewall performs packet filtering action a device takes to selectively control the flow of data to and from a network. Packet filters allow or block packets, usually while routing them from one network to another (most often from the Internet to an internal network, and vice versa). To accomplish packet filtering, a set of rules is set up to specify what types of packets (e.g., those to or from a particular IP address or port) are to be allowed and shat types are to be blocked. Packet filtering may occur in a router, in a bridge, or on an individual host. It is sometimes known as screening.

To provide more scrutinized security protection by using a firewall, it is desirable to keep tracking of network connections. To this end, stateful inspection takes the basic principles of packet filtering and adds the concept of packet transmission history, so that the Firewall considers the packets in the context of previous packets belonging to the same networking session (usually at the transport layer). A firewall can perform stateful inspection is called Stateful Firewall.

For example, a stateful firewall records when it sees a TCP SYN packet in an internal table, and in many implementations will only allow TCP packets that match an existing conversation by tracking the TCP connection and sequence numbers to be forwarded to the network. Using stateful firewall, it is possible to build up firewall rules for protocols, which cannot be properly controlled by packet filtering (e.g., UDP-based protocols). The drawback is that the stateful firewall's implementation is necessarily more complex and therefore more likely to be buggy. Moreover, it also requires a firewall device with more memory and a more powerful CPU, etc., for a given traffic load, as information has to be stored about each and every traffic flow seen over a period of time. In Table 4.2, a comparison among a few firewall technologies is provided.

In firewall terminology, a Bastion host is a special purpose computer on a network specifically designed and configured to withstand attacks. The computer generally hosts a single application, for example a proxy server, and all other services are removed or limited to reduce the threat to the computer. A Bastion host is usually considered as an application-layer gateway that is

TABLE 4.2

Comparison of Firewall Technologies

Firewall Capability	Packet Filters	Application-Layer Gateways	Stateful Inspection
Communication information	Partial	Partial	Yes
Communication-derived state	No	Partial	Yes
Application-derived state	No	Yes	Yes
Information manipulation	Partial	Yes	Yes

configured to inspect a specific application and it can check the correctness of application logics and application data content. It is hardened in this manner primarily due to its location and purpose, which is either on the outside of a firewall or in a DMZ and usually involves access from untrusted networks or computers.

An Example of Screened Subnet Firewall Architecture

Here, a screened subnet firewall architecture is illustrated. Figure 4.3 shows an example firewall configuration that uses the screened subnet architecture.

In this example, the perimeter network is another layer of security, an additional network between the external network and your protected internal network. If an attacker successfully breaks into the outer reaches of your firewall, the perimeter net offers an additional layer of protection between that attacker and your internal systems.

With a perimeter network (or DMZ), if someone breaks into a bastion host on the perimeter net, he will be able to snoop only on traffic on that net. All the traffic on the perimeter net should be either to or from the bastion host, or to or from the Internet. Because no strictly internal traffic (that is, traffic between two internal hosts, which is presumably sensitive or proprietary) passes over the perimeter net, internal traffic will be safe from prying eyes if the bastion host is compromised.

A bastion host is attached to the perimeter network and this host is the main point of contact for incoming connections from the outside world; for examples: for incoming email (i.e., SMTP) sessions to deliver electronic mail to the site; for incoming FTP connections to the site's anonymous FTP server; and for incoming domain name service (DNS) queries about the site. By

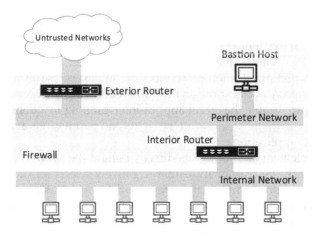

FIGURE 4.3
Traditional screened subnet architecture (using two routers).

setting up proxy servers to run on the bastion host to allow internal clients to access external network indirectly. We can also set up packet filtering to allow the internal clients to talk to the proxy servers on the bastion host and vice versa, but to prohibit direct communications between internal clients and the outside world.

The interior router (sometimes called the choke router in firewalls literature) protects the internal network both from the Internet and from the perimeter net. The interior router does most of the packet filtering for the firewall. It allows selected services outbound from the internal net to the Internet.

The exterior router (sometimes called the access router in firewalls literature) protects both the perimeter net and the internal net from the Internet. In practice, exterior routers tend to allow almost anything outbound from the perimeter net, and they generally do very little packet filtering. The packet filtering rules to protect internal machines would need to be essentially the same on both the interior router and the exterior router; if there is an error in the rules that allows access to an attacker, the error will probably be present on both routers.

The example presented in Figure 4.3 illustrated the mostly popular firewall setup for a private network connected to the Internet, where the trusted and untrusted domain boundary is clearly defined. However, with the fast-developed virtualization technology in both enterprise and datacenter networking environments, the boundary is not able to be defined clearly. As a result, the trusted domain boundary can be shrunk at the interface level, and a more scrutinized firewall system, e.g., distributed firewall, is required to provide protection on microsegmented networking systems, which will be described in Chapter 6.

4.3.2 Intrusion Prevention

An Intrusion Prevention System (IPS) is a network security/threat prevention technology that examines network traffic flows to detect and prevent vulnerability exploits. Vulnerability exploits usually come in the form of malicious inputs to a target application or service that attackers use to interrupt and gain control of an application or machine. Following a successful exploit, the attacker can disable the target application (resulting in a denial-of-service state) or can potentially access to all the rights and permissions available to the compromised application.

An IPS implementation often sits directly behind the firewall and provides a complementary layer of analysis that negatively selects for dangerous content. Unlike an Intrusion Detection System (IDS), which is a passive system that scans traffic and reports back on vulnerability exploits, the IPS is placed inline (in the direct communication path between source and destination), actively analyzing and taking automated actions on all traffic flows that enter the network. Specifically, these actions include:

- sending an alarm to the administrator (as would be seen in an IDS),
- dropping the malicious packets,
- blocking traffic from the source address, and
- resetting the connection.

As an inline security component, the IPS must work efficiently to avoid degrading network performance. It must also work fast because exploits can happen in near real-time. The IPS must also detect and respond accurately, so as to eliminate threats and false positives (legitimate packets misread as threats).

The IPS has a number of detection methods (which is also applied for IDS) for finding exploits, but signature-based detection and statistical anomaly-based detection are the two dominant mechanisms. Signature-based detection is based on a dictionary of uniquely identifiable patterns (or signatures) in the code of each exploit. As an exploit is discovered, its signature is recorded and stored in a continuously growing dictionary of signatures. Signature detection for IPS breaks down into two types:

1. Exploit-facing signatures identify individual exploits by triggering on the unique patterns of a particular exploit attempt. The IPS can identify specific exploits by finding a match with an exploit-facing signature in the traffic stream.

2. Vulnerability-facing signatures are broader signatures that target the underlying vulnerability in the system that is being targeted. These signatures allow networks to be protected from variants of an exploit that may not have been directly observed in the wild, but also raise the risk of false positives.

In addition to signature detection, there is another group of detection called statistical anomaly detection which takes samples of network traffic at random and compares them to a pre-calculated baseline performance level. When the sample of network traffic activity is outside the parameters of baseline performance, the IPS takes action to handle the situation.

4.4 Detection and Monitoring

4.4.1 Intrusion Detection

An Intrusion Detection System (IDS) is a network security technology originally built for detecting vulnerability exploits against a target application or computer. As we described previously, IPS extends IDS solutions by adding the ability to block threats in addition to detecting them and has become the dominant deployment option for IDS/IPS technologies. However, an IDS needs only to detect threats and as such is placed out-of-band on the network infrastructure, meaning that it is not in the true real-time communication path between the sender and receiver of information. Rather, IDS solutions will often take advantage of a TAP or SPAN port to analyze a copy of the inline traffic stream (and thus ensuring that IDS does not impact inline network performance).

IDS was originally developed this way because at the time the depth of analysis required for intrusion detection could not be performed at a speed that could keep pace with components on the direct communications path of the network infrastructure.

As explained, the IDS is also a listen-only device, i.e., a sniffer of network traffic. The IDS monitors traffic and reports its results to an administrator, but cannot automatically take action to prevent a detected exploit from taking over the system. Attackers are capable of exploiting vulnerabilities very quickly once they enter the network, rendering the IDS an inadequate deployment for prevention device.

In Table 4.3, we summarize the differences in technology intrinsic to IPS and the IDS deployment.

4.4.2 Logging

Logs are a critical part of a secure system; they give you deep insights about occurred events for later security analysis. Broadly speaking, we can categorize logging systems into three groups: (1) host-based log, (2) centralized log, and (3) decentralized log. The host-based logging solution stores logs on individual hosts. This approach minimizes the network traffic by transferring logs into a centralized log server; however, it incurs significant management overhead to retrieve logging data from individual hosts. Moreover, the

TABLE 4.3

Comparison of IPS and IDS

	IPS	IDS
Placement in Network Infrastructure	Part of the direct line of communication (inline)	Outside direct line of communication (out-of-band)
System Type	Active (monitor & automatically defend) and/or passive	Passive (monitor & notify)
Detection Mechanisms	1. Statistical anomaly based detection 2. Signature detection • Exploit-facing signatures • Vulnerability-facing signatures	1. Signature detection: • Exploit-facing signatures

logging data is vulnerable to host-based attacks, in which attackers may compromise a host and thus modify logged data.

For centralized logging approach, individual hosts or networking devices send their logs to a centralized logging service for log management and analysis. Using the centralized approach, log analysis model can be effectively deployed. It requires the centralized logging service to be robust against network attacks to avoid single point failure. For decentralized logging services, they are usually built on centralized logging solutions, in which multiple centralized-logging systems can be established to handle specific applications to address the scalability issues. Moreover, isolating different application logs, it will be easier to set up log analysis models with the application-focused log data.

For a given large datacenter networking environment, providing a centralized logging service, we usually require to establish four interdependent services, namely: collect logs, transport, store, analyze, and alerting.

Log Collection

All the applications create logs in different ways, some applications log through system logs (e.g., *syslog*) and other logs directly in files. When establishing a typical web application running on a Linux server, there will be a dozen more log files in */var/log* and also a few application-specific logs in the home directories and other locations. Basically, there will be logs generated by different applications at a different place.

There are two basic approaches to collect logs:

- Use the replication approach, where files are replicated to a central server on a fixed schedule. We can set up a *cron* job that will replicate log files on Linux server to the central log server. The *cron* job usually introduces delay due to the scheduled time intervals between two consecutive *cron* jobs.

- Use direct remote logging protocol to send log data when generated from the system without a delay. For example, we can configure syslog to send a new logged event to a remote logging server through a TCP port.

Transport

There are many frameworks available to transport log data. One way is directly plug input sources and framework can start collecting logs and another way is to send log data via API. Application code is written to log directly to these sources; it reduces latency and improves reliability. Some popular remote logging applications include: Logstash and Fluntd– open source log collector that are written in Ruby– and Flume– an open source log collector that is written in Java. These frameworks provide input sources but also support natively tailing files and transporting them reliably.

To log data via APIs, which is generally a more preferred way to log data to a central application, these are several frameworks that can be used; for examples, *Scribe*– an open source software by Facebook that is written in *C++* – *nsq*– an open source that is written in *Go*– and *Kafka*– an open source software by Apache that is written in Java.

Storage

To store logs, the storage system should be highly scalable, as the data will keep on growing and it should be able to handle the growth over time. Several factors should be taken into consideration:

- *Time – for how long should logged data be stored?* When logs are for long-term and do not require immediate analysis, they can be archived and saved on S3 or AWS Glacier, as they provide a relatively low cost for a large amount of data. When logs are for short-term, e.g., a few days or months of logs, we can use distributed storage systems like *Cassandra*, *MongoDB*, *HDFS* or *ElasticSearch*, etc. For ephemeral data, e.g., just a few hours, we can use *Redis* for example.

- *Volume – how huge would the logged data be?* Google and Facebook create a much larger volume of data in a day compared to a week's data of a simple *NodeJs* application. The storage system should be highly scalable and scale horizontally as log data increases.

- *Access – how will you access the logs?* Some storage systems are not suitable for real-time analysis, for example AWS Glacier can take hours to load a file. *AWS Glacier* or *Tape Backup* does not work if we need to access data for troubleshooting analysis. *ElasticSearch* or *HDFS* is a good choice for interactive data analysis and working with raw data more effectively.

Analysis

Traditional parallel data computing models such as *Hadoop* and *SPARK* can be used for log data analysis. However, the user needs to build the analysis algorithms. There are many tools specially designed for log analysis, if a UI for analysis is required, we can parse all the data in *ElasticSearch* and use *Kibana* or *Greylog2* to query and inspect the data. Moreover, *Grafana* and *Kibana* can be used to show real-time data analytics.

Alerting

Logs are very useful for troubleshooting errors and alert system administrators when critical events are detected. Using the alerting built in the logging application system, it will send an email or notify us then to have someone keep watching logs for any changes. There are many error reporting tools available; we can use *Sentry, Honeybadger,* or *Riemann*.

4.5 Network Security Assessment

A network security assessment will help us determine the steps we need to take to prepare ourselves, our organization, and our network for the threats of today and tomorrow. It must be a comprehensive evaluation by considering various attack scenarios based on attack models (from the attacker's perspectives) or threat models (from the defender's perspectives). An assessment should also consider the layered security infrastructure and provide a measurement approach in each step of the cyber killer chain to prevent or mitigate attacks at its earlier stage. Here are a few ideas to be considered for network security assessment:

- *Assess the vulnerabilities of networks, applications, and other IT resources.* Document and analyze the entire IT infrastructure to find the weaknesses and potential issues.

- *Conduct comprehensive scanning of ports, vectors, and protocols.* Conduct a comprehensive scan of all ports on the network to identify the IT equivalent of open windows and unlocked doors. The most common malicious network scans search for vulnerabilities in a standard range of 300 ports on a network where the most common vulnerabilities are found. However, we may have over 60,000 ports on our network that can be suspect.

- *Understand how your network interacts with outside parties.* Try to access the network as an outsider to inspect what network requests in terms of information and how easily it can be satisfied.

- *Probe internal network weaknesses.* Assess interaction with internal networks. Unfortunately, we cannot assume that all threats will originate outside the network. Internal people can pose a threat too.

- *Review wireless nets, including Wi-Fi, Bluetooth, RFID, and rogue devices.* Wireless nets, rogue devices, and removable media all present vulnerabilities. For example, a hacker may use baits, such as leaving a USB flash drive containing malicious code in the lobby, and then someone will likely pick it up and pop it into a system on the network to see what is on it.

- *Assess and educate employees about social engineering attacks.* This includes policies around behavior such as using social media or picking up flash drives left lying around, etc.

Summary

In this chapter, we provide a brief introduction of several important concepts of computer network security including attack/threat models, layers of security, preventive security defense, intrusion detection and monitoring services, and security assessment. The presented materials cannot cover all aspects of computer network security; however, it can provide a basic understanding of important security services for network security. Unlike data communications of the past, today's networks consist of numerous devices that handle the data as it passes from the sender to the receiver. However, security concerns are frequently raised in circumstances where interconnected computers use a network controlled by many entities or organizations. A comprehensive network security solution should examine various network protocols, focusing on vulnerabilities, exploits, attacks, and methods to mitigate an attack.

In order to have a better understanding of the security aspect of a network, readers who have difficulty to interpret the content presented in this chapter or who have little knowledge of computer network security may refer to other well-proposed network security books to enhance the security background to move forward to the advanced topics presented in this book.

5

SDN and NFV Security

Network Functions Virtualization (NFV) has emerged as a technology to provide a virtualized implementation of hardware-based equipment such as firewall, routers, and Intrusion Detection System (IDS). Virtual Network Functions (VNFs) can be realized through virtual machines (VMs) or containers running on top of the physical server of cloud computing infrastructure.

SDN acts as enabling technology for NFV. Despite the great benefits offered by SDN and NFV, the security, privacy and trust management remain an important problem to be addressed. The architecture of SDN and NFV has been discussed in previous chapters. In this chapter, we discuss the security challenges faced by different components of SDN and NFV, some that are part of traditional network architecture, and some introduced because of the SDN/NFV framework that should be considered before deployment of SDN/NFV technologies in a cloud network or data-center.

We survey the threat model and security challenges in NFV in Section 5.1. Section 5.2 has been dedicated to the classification of NFV security from the perspective of intra- and inter-virtual network functions (VNF) design. We also introduce some of the defense mechanisms that are used in NFV to deal with current threat vectors. In Section 5.3, we consider SDN security threat vectors. Section 5.2.2 provides guidelines for the design of a secured SDN platform. Additionally, we discuss the threat vectors specific to the SDN data plane, SDN architecture, OpenFlow protocol and OpenFlow switching software in this section.

5.1 Introduction

5.1.1 An Overview of Security Challenges in NFV

The NFV consists of two main function blocks, i.e., NFV MANO, and NFVI. The security of NFVI requires ensuring security compliance with standard methods of authentication, encryption, authorization and policy enforcement to deal with both internal and external threats.

TABLE 5.1

NFV Threat Vectors

Threat Vector	Description	Impact
VNF Service Flooding	Attackers can flood the service or the network interface using attacks such as DNS lookup, multiple authentication failure attempts resulting in denial-of-service (DoS) in signaling plane and data plane.	Availability
Application Crashing	Attackers can send malformed packets to the services running in NFV environment and cause network service disruption (e.g., buffer overflow exploit).	Availability
Eavesdropping	Attackers can eavesdrop on sensitive data and control plane messages.	Confidentiality
Data Ex-filtration	Unauthorized access to sensitive data such as user profiles.	Confidentiality
Data and Traffic Modification	Attacker can perform Man-in-the-Middle (MITM) attack on the network traffic in transit, perform DNS redirection or modify sensitive data on network elements (NE).	Integrity
Control Network and Network Elements	The attacker can exploit protocol vulnerabilities or implementation flaws to compromise a network. Additionally, attackers can exploit vulnerabilities on the management interface to take control of NE.	Control

5.1.1.1 NFV Threat Vectors

Threats to the NFV network perimeter and core services can violate the service level agreements (SLAs), such as NFV data confidentiality and service availability. We analyze the threat vectors that can impact the NFV framework in Table 5.1.

In addition to these threat vectors, there can be insider threats where attacker inside the NFVI can make changes to the data on Network Element (NE) or make changes to the network configuration.

5.1.1.2 NFV Security Goals

To define the security perimeter and its scope in NFV, we need to identify the security goals in NFV environment at various levels of granularity. The European Telecommunication Standards Institute (ETSI) defines the following security goals at a high level in an NFV environment:

- Establish a secure baseline of guidance for NFV operation while highlighting optional measures that enhance security to be commensurate with risks to confidentiality, integrity, and availability (CIA).
- Define areas of consideration where security technologies, processes, and practices have different requirements than of non-NFV systems and operations.

- Supply the guidelines for the operational environment that supports and interfaces with NFV systems and operations but avoid redefining any security considerations that are not specific to NFV.

The VNFs in an NFV environment often possess sensitive data and the NFV administrator should take care of data authentication in NFV workloads. The sensitive data authentication can consist of passwords, tokens, Cryptographic Keys, private keys and documents containing sensitive data. Each VNF can be responsible for one or more functions and capabilities. Authorization for the use of these functions and capabilities should be performed using standard techniques, e.g., identity, trust, delegated or joint decision making and API security.

The security in NFV is not limited to managing the NFV network and endpoints, but secured mechanisms must be designed for the lifecycle management of VNFs. The VNF creation requires changes to networking, credentialing, license and configuration information. Guidelines for VNF creation using newly defined configuration or cloning from a previously created VNF must be in place. Well-defined and secured mechanisms should be utilized for VNF lifecycle operations such as VNF deletion, Workload Migration, VNF configuration and patch management.

5.2 NFV Security

5.2.1 NFV Security Classification

NFV security architecture can be considered from various perspectives. Security domains of NFVI can be classified into networking, Compute and Hypervisor domains as discussed by Yang et al. [287]. ETSI [88] classifies the security domain of NFV into intra-VNF security, i.e., security between the VNFs and extra-VNF security, i.e., security external to VNF.

5.2.1.1 Intra-VNF Security

VNFs communicating with each other directly have special security requirements, since communication path is not restricted to the network level. The characteristics of intra-VNF security include:

- Secured orchestration for and between the VNFs.
- Flows between VNFs are not often through layer 3 firewall or any other security policy enforcement point.
- Service chaining capabilities often need to be enforced if available.

- Requires security mechanism in intra-VNF communication and resiliency to attacks.
- Security and virtual appliances need to be configured to be part of the traffic flow.

5.2.1.2 Extra-VNF Security

The VNF security is dependent upon the security of physical infrastructure, external services, and environment. The key issues that need to be considered for the security of NFV environment because of the factors external to the VNFs are

- NFV deployment may span across several regulatory and jurisdiction domains, leading to multiple sets of Service Level Agreement (SLA) and Quality of Service (QoS) requirements. Extra-VNF security should have the ability to administer cross-border and domain requirements, e.g., Workload Migration from one public NFV tenant to a secured NFV tenant may impact the QoS or security of the destination NFV tenant.
- Authentication, authorization, and accounting across NFV domains, across a mix of domains, humans and system entities. For instance, one NFV deployment can have multiple administrative domains, e.g., a) NFVI, b) SDN, c) Orchestration, d) VNF Manager (VNFM), e) Service Network.

5.2.2 NFV Security Lifecycle

The VNF lifecycle as shown in the Figure 5.1(a) comprises five phases, i.e., VNF development, instantiation, operation, enhancement, and retirement. The security management processes 5.1(b) for NFV should be embedded into these phases of VNF lifecycle. The scope of NFV security comprises of NFV framework, hardware, software and service platform that supports NFV. We consider hardware platform to be following the required NFV security guidelines and consider security for hardware platform out of the scope of this chapter.

We can segment the NFV architecture into components that can be a direct or indirect target of attacks. The goal of the attack can range from reconnaissance, service degradation, service disruption to unauthorized access to critical information in the NFV framework. Reynaud et al. [228] have identified five critical assets in NFV framework that can be potential attack targets: (1) Virtual Network Functions, (2) Virtualization Layer, (3) Communication with and within NFV MANO, (4) NFV Manager and/or Orchestrator, (5) Virtualized Infrastructure Manager (VIM) as shown in Figure 5.2.

1. **Virtual Network Functions:** VNFs suffer from software vulnerabilities. They can be a source of an attack or target. The vulnerabilities

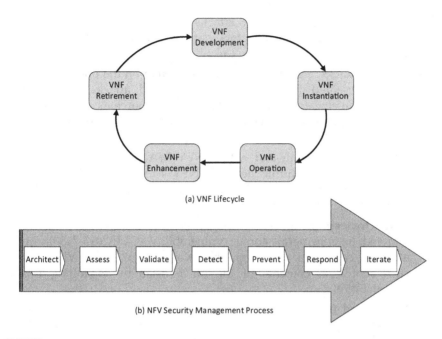

FIGURE 5.1
Virtual networking approaches enable different logical views of the underlying physical networks.

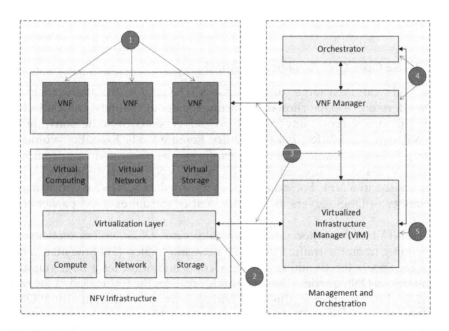

FIGURE 5.2
NFV targetable components.

like buffer overflow and DoS attacks against cloud and web-based services are typical threat vectors that can be caused by VNFs.

2. **Virtualization Layer:** The virtualization layer can be a target of many security attacks, e.g., malicious code execution on the physical host, Return Oriented Programming (ROP) based attacks, where an attacker can elevate the VM privilege, CPU resource monopolization attack, Data Theft and VM monitoring attacks as discussed by Riddle and Chung [229].

3. **Communication with and within NFV MANO:** The attacker can eavesdrop on the traffic between NFV MANO and NFVI. The attacker can perform a MITM attack, an attack vector targeting this particular communication channel.

4. **VNF Orchestrator and/or Manager:** NFV over OpenStack can be targeted to ephemeral storage vulnerability (CVE-2013-7130). An attacker can steal Cryptographic Keys from other VNFs or steal root disk contents of the other users by exploiting this vulnerability.

5. **Virtualized Infrastructure Manager (VIM):** Attacks can target the infrastructure manager in the NFV, e.g., Ruby vSphere console in VMWare vCenter Server suffers from privilege escalation vulnerability (CVE-2014-3790). This allows remote users to escape chroot jail and execute arbitrary code in the infrastructure domain.

5.2.3 Use Case: DNS Amplification Attack

The VNFs can be a target of DoS attacks. The goal of the attacker can be network resource exhaustion or impacting service availability. If the attacker can exploit a vulnerability present on old versions of some software of a VNF, e.g., CVE-2018-0794 (MS Office Remote Code Execution vulnerability). A huge volume of network traffic can be generated from compromised VNFs and directed towards other VNF present on the same Hypervisor in VNFI. For example, Figure 5.3 shows NFVI comprising of a number of DNS servers as a component of virtual evolved packet core (vEPC).

The NFVI orchestrator can spawn additional DNS servers on-demand depending upon the traffic load in the network. In Step (1) of the attack, the attacker spoofs the IP address of the victims and launches a large number of malicious DNS queries. The orchestrator realizes the traffic load in the network is above the normal threshold and spawns out additional virtual DNS (vDNS) VNFs in Step (2). Multiple recursive DNS servers in the network respond to the victim, and in-effect receive amplified DNS query responses - Step (3), which can ultimately result in service unavailability or disruption.

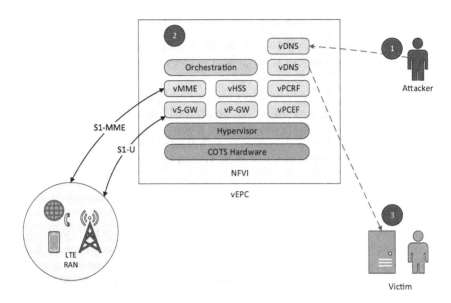

FIGURE 5.3
DNS amplification attack in NFV environment.

5.2.4 NFV Security Countermeasures

The ETSI NFV Industry Specification Group (ISG) and NFV Security Expert Group have identified some key areas of concern in NFV security and security best practices to deal with these security problems. In this section, we discuss NFV security countermeasures based on ETSI specification and state-of-the-art research works in the field on NFV security.

5.2.4.1 Topology Verification and Enforcement

The network topology and communication of data plane, control plane and management plane in NFV should be validated, ensuring the following specifications.

Data Plane

- Intra-host communication (the communication between VNFs on the same host).
- Inter-host communication (communication between VNFs on different Hypervisors).
- Communication path between VNFs and physical equipment.

Control and Management Plane

- The communication paths within the MANO system.

- Paths between MANO and the Virtualized Infrastructure.
- Paths between MANO and the hardware infrastructure.
- Paths between MANO and the managed VNFs.

The topology of these two networks must be validated individually as well as together. The topology validation can be divided into different levels to manage the complexity of the operation. For instance, the physical and logical topology (VLAN, GRE) of the underlying infrastructure can be checked first, followed by validation of ports of each virtual forwarding function in VNF environment.

5.2.4.2 Securing the Virtualization Platform

An important security assumption in NFV is that a VNF provider has to trust the virtualization platform on which various VNFs have been hosted. Additionally, the platform should also have some mechanism to ensure the trust in VNFs. One way of providing platform security is Secure Boot [78] technology. Secure boot can help maintain validation and assurance of Boot Integrity. There are several assurance factors that are part of Boot Integrity, including authenticity, configuration management, local attestation, certificates, Digital Signatures, etc.

A malicious attacker can tamper with the initial boot process of VNF to load malicious code during the VNF launch cycle. Secure boot can provide assurance that the code loaded in VNF execution environment is authentic, and has not been tampered with. Trusted boot process in coordination with VNF manager can provide validation during the VNF launch and installation stages.

5.2.4.3 Network and I/O Partitioning

One of the main purposes of virtualization is the isolation of VMs from crashes, loops, hangs, and security attacks from other VMs. The objective is hard to be realized when:

- Granularity at which network boundaries have been defined or resources have been allocated is too coarse.
- Use of workloads is highly variable.

There are various attack vectors that can target Hypervisor resources in an NFV environment, e.g., a) local storage attack can be mounted to fill up Hypervisor local storage with logs, b) remote connection attacks (remote control channel degradation).

The network resource is a critical network function. In addition to the malicious users, sometimes a large number of remote users can send a request to local resources, and it is hard to distinguish between the normal request from malicious traffic. An efficient QoS scheme can be used to ensure that critical tasks are given priority in case of high network demand. Additionally, the network must be partitioned into fine-grained segments to localize the threat only to the infected segment of the network.

Resource isolation is another mechanism to achieve fine-grained partitioning. Some methods to achieve isolation include a) physical segregation of hardware resources, b) rate-limiting the usage of resources VNF can reserve, c) dividing available resources between competing demands using efficient scheduling mechanism, e.g., round-robin or fair-queue bandwidth scheduling.

5.2.4.4 Authentication, Authorization, and Accounting

Reliable mechanisms to ensure the identity and accounting facilities at the network, and virtualization layer can be incorporated to achieve authentication, authorization and accounting (AAA).

Introduction of NFV can bring new security issues for AAA. The identity and accounting facilities span across two regions, i.e., network infrastructure layer (identifying the actual tenant), and network function layer (identifying the particular user), as shown in Figure 5.4. Some of the AAA issues that can occur in NFV framework include:

1. **Authentication:** Unauthenticated disclosure of user information at the layers that are not supposed to consume certain identity attributes.

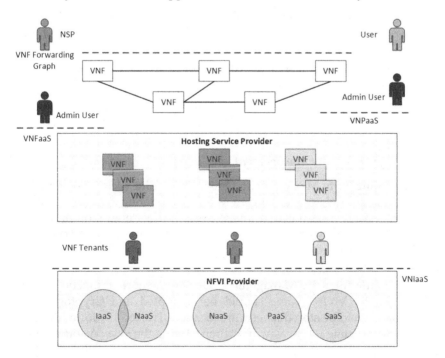

FIGURE 5.4
DNS amplification attack in NFV environment.

2. **Authorization:** Privilege escalation by wrapping unrelated identities not verifiable at a particular layer.

3. **Accounting:** Lack of accounting at different layers of network infrastructure, e.g., the granularity of tenant can allow an attacker to over-subscribe the allocated resources in NFVI.

A generalized AAA scheme is required to support identity and access management at each tenant and VNF level. The current AAA mechanisms assume there will be a single identity, policy decisions, enforcement points, and single accounting infrastructure. Achieving this in current NFV framework using mechanisms such as tunneling can introduce scalability and resiliency concerns. Although achievement of all possible objectives such as security enforcement, scalability, flexibility, manageability are difficult to achieve, it is important to find the appropriate combination supporting the trust framework. Some of the countermeasures include:

- Authentication of VNF images.
- Authentication of users requesting access to NFV MANO function blocks.
- Updates to authorized users and managers in the suspended/offline images.
- Authorization on interfaces/APIs between different function blocks.
- Support for real-time monitoring, logging, and reporting on SLAs, reliability, and performance.
- Traffic packet acquisition at full line rate and traffic classification and accounting per subscriber, per user and per application.
- Policy decision functions that raise alarm when a specific threshold has been reached according to the detected policies and traffic.

5.2.4.5 Dynamic State Management, and Integrity Protection

- **Dynamic State Management:** Online and offline security operations such as securely suspending a VM image, updating the access control lists (ACLs) in suspended images, secured live migration of VNFs. VNFC should be provisioned with the initial root of trust. All communication between the initial root of trust and VNFC should be strictly monitored.

- **Dynamic Integrity Management:** During the normal functioning of VNF, the VNF volume should be encrypted and Cryptographic Keys should be stored in a secure location. Trusted Platform Module (TPM) volume is one way of securely storing the keys. A software or hardware misconfiguration can cause VNF to crash, rendering the VNF in an unexpected state, which can cause security concerns. The crash events should be properly analyzed to ensure the integrity of VNF keys and passwords is maintained during the crash. The analysis of crash event should also consider external influences that should be mitigated to restore service. In case of crash events, the Hypervisor should also be properly configured to wipe out virtual volume disk to prevent it from unauthorized access.

5.3 SDN Security

SDN finds many applications in enterprise cloud and data-center network. The adoption of SDN can provide benefits in not only cloud management and orchestration but also cloud security. Thus, the security of SDN itself is quite an important area of research. The centralized design of SDN can introduce security challenges such as distributed denial-of-service (DDoS) attacks against the SDN controller. The SDN functional architecture can be divided into three layers, i.e., the application layer, the control layer and the data layer as discussed in previous chapters. Each layer can have multiple attack vectors. Additionally, the communication channel between layers, e.g., an application-control interface can be targeted to traffic modification and eavesdropping attacks.

5.3.1 SDN Security Classification

The relationships between SDN elements can introduce new vulnerabilities, which are absent in the traditional network. For instance, the use of transport layer security is optional in the OpenFlow network. The nature of the communication protocol can thus introduce security issues such as DoS, fraudulent flow rule insertion, and rule modification as discussed by Scott-Hayward et al. [241].

Figure 5.5 highlights different components in SDN: (1) application plane, (2) control plane and (3) data plane tier that can be subjected to attacks. For instance, there can be software vulnerabilities in SDN controllers (Opendaylight, ONOS, Floodlight). Additionally, the communication paths between three tiers, i.e., northbound APIs (4) and southbound APIs (5) can face security attacks. We discuss some of the attack vectors against targetable components in detail below:

- **Application Plane:** The applications developed for telemetry, orchestration and other SDN operations can have security vulnerabilities. All the security issues that can be present in a typical web application such as Cross Site Scripting (XSS), Cross Site Request Forgery (CSRF) also apply to SDN. The malicious/compromised applications can allow spread of attack in the entire network.

- **Control Plane:** The control plane consists of one or more controller, e.g., OpenDaylight, POX, ONOS and other applications and plugins

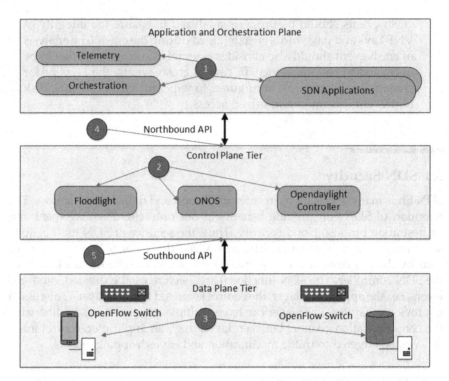

FIGURE 5.5
SDN targetable components.

for handling different kinds of protocols. The attacker can generate traffic from spoofed IP address and send a huge volume of traffic to the controller as discussed by Kalkan et al. [142]. The communication between the switch and the controller can be saturated using this attack, thus increasing service latency or in the worst case bringing down the controller.

- **Data Plane:** The attackers can poison the global view of the network, by forging the Link Layer Discovery Protocol (LLDP) packages. The attackers can also observe the delay in communication between the control plane and data plane applications using specially crafted packets. This can help in identification of controller application logic [96]. The attackers can also target the switches. The switch responsible for data plane flow rule updates often have limited memory and can be overflowed by generating a large number of flow rules.

- **Communication Channels:** The communication channel between switches and controllers (Southbound API), controllers and application plane tier (Northbound API) can be subjected to Man-in-the-Middle (MITM) attack as showcased by Romao et al. [230], ARP

TABLE 5.2

Security Issues Associated with Different Layers of SDN

Security Attack	SDN Layer Affected				
	App Layer	App-Ctl Intf	Ctl Layer	Ctl-Data Intf	Data Layer
Unauthorized Access					
Unauthorized Controller Access			✓	✓	✓
Unauthenticated Application	✓	✓	✓		
Data Leakage					
Flow Rule Discovery					✓
Forwarding Channel Discovery					✓
Data Modification					
Flow Rule Modification			✓	✓	✓
Malicious Applications					
Fraudulent Rule Insertion	✓	✓	✓		
Controller Hijacking			✓	✓	✓
Denial of Service					
Controller Switch Flooding			✓	✓	✓
Switch Flow Table Flooding					✓
Configuration Issues					
Lack of TLS			✓	✓	✓
Policy Enforcement Issues	✓	✓	✓		

Poisoning is one example of such security attacks. Other attacks showcased by authors that target the communication channel include eavesdropping traffic between hosts, and stealthily modifying the traffic between hosts.

Table 5.2 summarizes some of the security issues associated with different components of SDN that we described above.

5.3.1.1 SDN Security Threat Vectors

In this section, we discuss some key Threats Vectors (TVs) in SDN in detail, and analyze if a better SDN platform design can help in dealing with security threats intrinsic and extrinsic to the SDN.

- **TV1 Fake Traffic Flows:** Faulty devices or malicious users can use DoS attacks to target the TCAM (ternary content-addressable memory) switches in the SDN infrastructure, with the goal of exhausting the capacity of the TCAM switches. The problem can be mitigated by using a simple authentication mechanism, but if the attacker is able to compromise the application server consisting of details of users, an attacker can use the same authenticated ports and source MAC addresses to inject forged authorized flows into the network.

- **TV2 Switch Specific Vulnerabilities:** The switches present in the SDN environment can have vulnerabilities. For instance, a vulnerability in Juniper OS (CVE-2018-0019) SNMP MIB-II subagent daemon (mib2d) allows a remote network-based attacker to cause the mib2d process to crash resulting in a denial of service condition (DoS) for the SNMP subsystem. A switch can be used to slow down the traffic in SDN environment, deviate the network traffic to steal information, or can be used to insert forged traffic requests with the goal of overloading the controller or the neighboring switches.

- **TV3 Control Plane Communication Attack:** The control-data plane communication does not require the presence of TLS/SSL security. Even if Public Key Infrastructure (PKI) is present in an SDN environment, complete security is not guaranteed for the channel communication. Research works highlight security issues with TLS/SSL [113]. A compromised Certificate Authority (CA), vulnerable application can lead to an attacker gaining access in control plane channel of the SDN. The attacker can launch DDoS by using switches that are controlled by the control plane.

- **TV4 Controller Vulnerabilities:** The controller is the most important component in the SDN environment. A compromised controller can bring down the entire network. For example, an old version of SDN controller ONOS suffers from remote denial of service attack (CVE-2015-7516). The attacker can cause NULL pointer dereference and switch disconnect by sending two Ethernet frames with ether_type Jumbo Frame (0x8870) to ONOS controller v1.5.0. A combination of signature-based intrusion detection tools may not be able to find the exact combination of events that triggered a particular behavior and deem it malicious or benign.

- **TV5 Lack of Trust between Controller and Management Applications:** Controller and management plane applications lack a built-in mechanism to establish trust. The certificate creation and trust verification between network devices in the SDN environment can be different from the trust framework between normal applications.

5.3.2 Design of Secure and Dependable SDN Platform

A secure and dependable SDN architecture as shown in Figure 5.6, having features such as fault-tolerance, self-healing, trusted framework and dynamic service provisioning capabilities can be used to deal with threat vectors discussed in the previous subsection. In this section, we discuss each of the

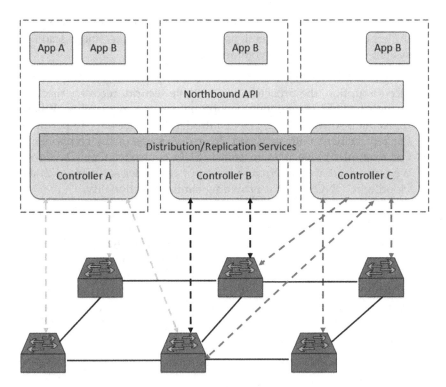

FIGURE 5.6
Design of secure and dependable SDN.

security mechanisms that can be embedded into the design of the SDN framework.

1. **Replication:** Application and controller replication can help in dealing with cases of controller or application failures due to a high volume of traffic or software vulnerabilities. As shown in Figure 5.5, there are three versions of the SDN controller providing replication. Additionally, application B has been replicated on each controller. This approach can help in dealing with both hardware and software failure issues (accidental or malicious). Another advantage of replication is the isolation of malicious application while keeping the service consistency.

2. **Diversity:** The utilization of only one kind of software or operating system makes it easier for attackers to exploit a target. Diversity improves the robustness and intrusion tolerance. As discussed by Garcia et al. [97], utilization of a diverse set of OS makes a system less susceptible to intrusions. Diversity helps in avoiding the common faults and vulnerabilities since there are only a few intersecting vulnerabilities among

diverse software or OS. In SDN management plane, use of diverse controllers can help reduce lateral movement of an attacker and cascading system failures caused by common vulnerabilities.

3. **Automated Recovery:** In the case of security attacks, leading to service disruption, the proactive and reactive security recovery mechanisms can help in maintaining optimal service availability. When replacing a software, e.g., SDN controller, it is necessary to perform the replacement with new and diverse versions of the component. For example, if we plan to switch SDN controller OpenDaylight, we can consider an alternate version of controller software such as Floodlight, ONOS or Ryu providing similar functionality.

4. **Dynamic Device Association:** The association between the controller and devices such as OpenFlow switch should be dynamic in nature. For instance, if one instance of the controller fails, the switch should be able to dynamically associate with the backup controller in a secured fashion (proper authentication mechanism to detect good controller from malicious controller software). Dynamic Device association feature helps in dealing with faults (crash or Byzantine). Other advantages include load balancing feature provided by diverse controllers (reduced service latency).

5. **Controller-Switch Trust:** A trust establishment mechanism between the controller and switch is important to deal with cases of fake flows being inserted by malicious switches. The controller can in basic trust establishment scenario maintain a whitelist of switch devices that are allowed to send control plane specific messages to the controller. In a more complex scenario, Public Key Infrastructure (PKI) can be used to establish trust between the control plane and data plane devices. The behavior devices controlled by the controller can also be used to create a trust framework. The devices showcasing anomalous behavior can be put in quarantine mode by the controller.

6. **Controller-App Plane Trust:** The software components change behavior because of change in the environment. Additionally, the software aging can introduce security vulnerabilities. Controller and application plane components should use autonomic trust management mechanisms based on mutual-trust and delegated trust (3rd part such as the Certificate Authority to establish trust). The controller can utilize autonomic trust management for component-based software systems as discussed by Zheng and Prehofer [285]. Qualitative metrics such as confidentiality, integrity, and availability can also be leveraged to establish the trustworthiness of an application in the SDN framework.

7. **Security Domains:** Security domains help in segmenting the network into different levels of trust, and containment of the threats to only the affected section in the SDN framework. A security domain-based

TABLE 5.3

SDN Design Solution for Dealing with Threat Vectors

SDN Security Solution	Threat Vector
Replication	TV1, TV4, TV5
Diversity	TV3, TV4
Automated Recovery	TV2, TV4
Dynamic Device Association	TV3, TV4
Controller-Switch Trust	TV1, TV2, TV3
Controller-App Plane Trust	TV4, TV5
Security Domains	TV4, TV5

isolation can be incorporated to provide defense-in-depth for SDN environment. For example, the web-server application on one physical server should only interact with database back-end applications, and not any other application running in the same network. A whitelist-based security policy composition with appropriate policy conflict checking mechanism can be utilized to achieve such security objectives. We discuss the segmentation policy creation, its key benefits, and real-world applications in detail in the next chapter.

Table 5.3 summarizes the security solutions that we discussed in this subsection and threat vectors TV1-5, that can be mitigated using these mechanisms. An important consideration while deploying the desired security solution or combination of solutions is cost-benefit analysis (delay introduced in the network, CPU/resource utilization) of these solutions in isolation as well as together.

5.3.3 SDN Data Plane Attacks and Countermeasures

5.3.3.1 SDN Data Plane Attacks

The communication between the SDN data plane and control plane opens the doors for highly programmable application development; it also introduces the possibility of new types of threat vectors in SDN data plane. The data plane attacks can dysfunction the anonymous communication in OpenFlow networks [292]. The attacker can also perform reconnaissance on the traffic channel between SDN data plane devices and controller to identify relevant information about controller software, e.g., version and type of controller (python or java application). This information can be utilized by the attacker to perform more targeted attacks against control plane software. Some key attacks that find origin in the SDN data plane include:

1. **Side-Channel Attacks:** The attacker can observe the processing time of the control plane in order to learn the network configuration as

discussed by Sonchak et al. [256]. The attacker crafts probe requests corresponding to different layers of network protocol stack, e.g., ARP requests for MAC layer and TTL probe message for the IP layer. The requests can be sent to the controller along with some baseline traffic requests with a known response. By observing the response time and content difference between baseline traffic and the probe traffic, the attacker can observe a version of OpenFlow, the size of switch flow table, network and host communication records, network monitoring policies, and the version of the controller software.

2. **Denial-of-Service (DoS):** The devices send a connection request to the switching software in the data plane. If the switch consists of flow rule entry corresponding to the traffic pattern, traffic is forwarded out of the specific switch port. If the entry is missing (table-miss packets) the request is sent to the controller. A class of DoS attacks - data to control plane saturation attacks as discussed by Gao et al. [96] can forge the OpenFlow fields with random values, that will lead to table-miss event in the switch. When a large volume of forged table-miss flows is sent to the controller as packet_in entries. The controller can be saturated since these packet_in messages will consume a large amount of switch-controller bandwidth and controller resources (CPU, memory). In case the controller decides to insert the flow entries in the switch, the TCAM table limit of the switch may be reached, which will prevent the legitimate traffic flows from being inserted in the switch flow table.

3. **Topology Poisoning Attacks:** This is a two-stage data plane attack, which utilizes forged control plane packets (LLDP packets) as a starting point. In the first attack, the attacker captures the OpenFlow LLDP packets and filters out the LLDP syntax. In the second step, the attacker sends the forged LLDP packets to the controller or replays them to other hosts in the network, in order to trigger the response from the connected switch to the controller. This can help an attacker in establishing a previously non-existent link between the switches. The attacker can utilize the modified topology to his advantage and launch MITM or a variation of DoS attack in the network.

5.3.3.2 SDN Data Plane Attack Countermeasures

1. **Side-Channel Countermeasure** The side-channel attacks rely on response time pattern, so a disruption in time frequency can help counter side-channel attacks. Sonchak et al. [256] propose timeout proxy on the data plane to normalize control plane delay. When the control plane fails to respond within a specified time duration, the timeout proxy sends default forwarding instruction to the request. The proxy reduces the long delay time of some long response packets

to avoid side-channel attacks. The duration of response can also be randomized in order to counter the side-channel attacks.

2. **DoS Countermeasure** DoS attacks such as SYN flood can be inspected for existence of valid source and destination addresses. AvantGuard extends the hardware of OpenFlow switches, and adds TCP proxy to send SYN-ACK as a reply for the TCP-based data to control plane starvation attacks. Another approach is to perform statistical analysis and flow classification to distinguish attack traffic from the benign traffic which can help in the detection and prevention of DoS attacks.

3. **Topology Poisoning Countermeasure** The type of neighbor devices connected to OpenFlow switch can help in dealing with LLDP-packet-based Topology Poisoning Attacks as showcased by TopoGuard [119]. The control plane can identify neighbor device using packet hop distance, and other packet statistics. If a port first receives an LLDP packet, the neighboring device can be regarded as a switch. On the other hand, if a packet from first hop host is received, the neighboring device is regarded as host. The dynamic monitoring and probing can help in the reconstruction of neighbor topology. The drawback of this approach is that it may allow attackers to forge neighbor device transfer from host to switch. Gao et al. [96] propose more a granular classification of devices (host/switch/ any/untested) to deal with the problem of topology poisoning faced by TopoGuard.

5.3.4 SDN-Specific Security Challenges

In addition to the threat vectors discussed in previous subsections, there are some security issues specific to SDN that are not inherently present in traditional networks. In this subsection, we highlight these security challenges and best practices to avoid them in SDN.

5.3.4.1 Programmablity

SDN offers programmatic features to the clients who belong to different business entities and organizations. Traditional business entities follow a closed domain, administrative model. Thus the SDN business model makes it necessary to protect system integrity, open interfaces and 3rd-party data across multiple administrative and business domains.

- **Traffic and Resource Isolation:** The business management and real-time control information of one application need to be fully isolated from other applications. Traffic and resource isolation across tenants must be ensured in the SDN environment. The SLA requirements and private addressing scheme induced dynamic interactions may create a need for more fine-grained isolation.

- **Trust between third-party applications and controller:** Authentication and authorization mechanisms should be enforced at the point of application registration to the controller in order to limit controller exposure.

5.3.4.2 Integration with Legacy Protocols

The advent of SDN solved some of the technical and process deficiencies in the legacy protocols. However, retrofitting of the security capabilities into existing technologies, e.g., DNS, BGP may not be straightforward. It is critical to inspect compatibility of legacy protocols before incorporation into SDN.

5.3.4.3 Cross-Domain Connection

SDN infrastructure allows connectivity across different physical servers, clusters and data centers. Each security domain can be under the control of one or many controllers. An appropriate mechanism for establishing a trust relationship between controllers should be present in SDN design. The trust framework should have the ability to prevent abuse and capability of establishing a secure channel.

5.3.5 OpenFlow Protocol and OpenFlow Switch Security Analysis

5.3.5.1 Attack Model

Actors

The threats against OpenFlow protocol can be internally initiated or externally initiated. A trusted insider can try privilege escalation to modify the implementation of OpenFlow protocol or perform unauthorized access request on the OpenFlow related reference data. On the other hand, an external attacker can control the dataplane devices directly attached to the OpenFlow switches, and try to generate malicious traffic request aimed at disrupting the communication or gaining privileges to OpenFlow devices remotely.

Attack Vectors

The following attack vectors can be employed by external and internal attackers who aim to target OpenFlow components:

- Passive eavesdropping on the data/control plane messages. This may help the attacker gain necessary information for subsequent attacks.

- Replay attacks in SDN network with non-authentic data/control messages, Man-in-the-Middle (MITM) attack, DoS/DDoS attacks or side-channel attacks.

Target/Goal

The attacker may aim at obtaining OpenFlow protocol assets/properties:

- Sensitive information in protocol messages.
- Tenant, network topology, SDN network availability or performance related information.
- Reference data on devices implementing OpenFlow switch flow table entries.
- Data and resource information of control and dataplane (e.g., bandwidth, latency, flow timeout duration).

5.3.5.2 Protocol-Specific Analysis

The protocol specific analysis should consider following entities, components and subcomponents as shown in Table 5.4. We discuss the candidate security countermeasures to deal with attacks against OpenFlow protocol, OpenFlow switch and associated components highlighted in Table 5.5. The analysis assumes that each OpenFlow switch can be connected to one or more controllers within the trust boundary of cloud service provider. Also, TLS security can be employed between switch and controller to deal with message tampering and to perform mutual authentication.

TABLE 5.4

OpenFlow Protocol Analysis Breakdown

Entity	Component	Sub-components/Scenarios
Switch	Ports	• Physical Ports • Logical Ports • Reserved Ports
	Tables	• Counters
OpenFlow Channel and Control Channel	Channel Connections	• Connection Setup • Encryption • Multiple Controllers • Auxiliary Connections

TABLE 5.5

OpenFlow Component Security Issues and Candidate Countermeasures

Component	Security Issue	Candidate Countermeasure
Physical Port	Fake physical port may be inserted or changed in order to perform traffic analysis leading to network attack.	Enable link state monitoring and network change tracking capability in SDN controller.
Logical Ports	Port tunnel ID missing in port statistic messages.	Enable tunnel ID checking feature in the controller.
Reserved Ports	Controller unable to collect statistical information on reserved ports.	Enable APIs to allow the controller to query reserved ports.
Counters	Counter roll-back out of control	Ensure controller, flow table synchronization.
Connection Setup	TLS protection for TCP header missing or information to manage Cryptographic Keys and certificates missing.	Mechanisms providing TCP-AO for header protection, and switch management protocol for key and certificate management should be present.
Encryption	Authentication for message communication not present.	Support for multiple types of authentication and encryption protocols should be incorporated.
Multiple Controller	Security policy conflict between controllers, malicious controller attempting unauthorized access in the network	Mutual authentication and synchronization should be employed across controllers. Role-based authentication for each controller. Secure communication between the master controller and switches.
Auxiliary Connections	Lack of verification mechanisms against invalid DPID.	Alert mechanism in the controller when invalid DPID sends across a packet. Use different authentication for auxiliary and main connections.

Summary

The SDN and NFV platforms suffer from many threat vectors, some of which are introduced by weak authentication and authorization mechanisms, others because of the SDN/NFV design. Consideration of each threat vector in isolation is important for creating a secured cloud networking environment managed by SDN/NFV. This chapter explored security issues affecting confidentiality, integrity, and availability of SDN/NFV. The security design goals and best practices, security countermeasures have described in detail for NFV, SDN data plane, SDN control plane and OpenFlow protocol. The secured architecture design depends on many other factors apart from the mechanisms described in this chapter, such as latency and throughput impact because of a particular secured configuration. These factors, however, are beyond the scope of this chapter and should be considered before adopting the recommendations provided as a part of this chapter.

Summary

Part II

Advanced Topics on Software-Defined and Virtual Network Security

In this part, we will provide advanced topics for virtual network security and particularly several important topics such as Moving Target Defense (MTD), attack graph and attack tree based security analysis approaches, security service function chaining, security policy management, and machine-learning based attack analysis models are presented. Moreover, the second part can be used by senior undergraduate students or graduate students as a reference book to work on their thesis or research-related projects. Before moving forward, several important and highly related terms need to be clearly understood.

What is *Network Security*?

Network security is the activity designed to protect the usability and integrity of the network and data. The security is not limited to the network edge or core, it includes hardware, software and human aspect within and outside an organization. While there is no silver bullet to achieve complete network security, the network security should at minimum ensure confidentiality, integrity and availability (CIA) of the data at rest (i.e., present inside the corporate environment) in form of critical databases, active directory information of users in the organization, files containing sensitive information, and data in transit, i.e., data should be securely transmitted to the trusted party with appropriate authentication and authorization.

There are various tools and techniques in traditional network security that use reactive threat mitigation techniques such as Network Firewall, Web Application Firewall, and Intrusion Detection System (IDS), to name a few. The goal of this part of the book is to introduce the latest proactive security solutions that have gained popularity in the industry as well as academic research, with emphasis on Software-Defined Networking and its role in adoption of these security solutions.

What is *Software-Defined Security?*

Software-Defined Data Centers (SDDCs) are able to allocate compute, storage and networking resources dynamically. While the modern-era data centers are evolving to become software defined, we lag behind in the security of virtual infrastructure. Current information security methods are too rigid and static to support the rapidly changing digital business. SDN provides next generation information security services that enforce proactive monitoring, analysis, detection and prevention of threats in the Virtualized Infrastructure. The SDN+next generation smart information security is together referred to as Software-Defined Security (SDS). The security infrastructure in SDS will be able to adapt to changes in network infrastructure and application services. In the long term, the adaptive security infrastructure will be driven by software-defined models, providing protection against emerging threats.

Some key features of SDS as per Gartner, Inc. are

- Location independent security for information and workloads.
- Aligning security controls to the risk profiles of what they are protecting.
- Enabling policy driven and automated security orchestration and management.
- Removing the time- and error-prone human middleware via higher level of automation.
- Enabling information security professionals to focus on policies and detecting advanced threats using programmable security components.
- Enabling security to scale and protect dynamic cloud-based workloads.
- Enabling security to move at the speed of digital business.

Distributed Security and Microsegmentation

Traditionally security is provisioned in a centralized way, distributed security is a method of breaking traditional data center and cloud network into logical elements and managing each element separately. Such architecture allows enforcement of granular security, i.e., security at the granularity of each network, subnet and application workload. Microsegmentation is one such approach towards a decentralized security framework. Network microsegmentation provides software-level abstraction to the subnetwork traffic control, making security management architecture simpler in complex SDDC with fluctuating workloads and applications. Another key benefit provided by microsegmentation is the prevention of lateral movement of the attacker by introducing zero-trust proof architecture.

Proactive Security

Proactive security refers to a new paradigm of security management, where instead of defending the network infrastructure by detecting, preventing, tracking and remediation of threats, attack surface is changed over time. This creates an asymmetric disadvantage for the attackers, and levels the playing field between the attacker and defender. MTD is a term that has been coined to generalize different proactive security mechanisms such as introducing diversity into network topology, OS, software, or randomizing the memory layout so that the attack propagation becomes difficult for the attacker.

Security Policy Management

A large organization can have different groups, each responsible for managing a small portion of the network. Often there is high interdependence between the access control policies of different groups. The security policy management incorporates identification and correction of policy conflicts between different security policies in an automated fashion.

Attack Representation Methods

The interaction between different applications and the impact on the attack surface by the dynamics of applications and workloads changing constantly in a cloud network needs to be expressed in an intuitive fashion, especially for the network security administrators, so that they can make an informed decision, keeping the entire threat landscape of the organization into perspective. Attack Representation Methods (ARMs), such as attack graphs and attack trees, help in representing complex information in an easy-to-understand format. ARMs in combination with the security metrics, such as Common Vulnerability Scoring System (CVSS), allow network administrators to ask higher order logic questions, such as the likelihood of critical services at the core of networks being exploited, if the attacker starts from the vulnerable public facing website at the edge of the network, or return of investment (ROI) from the security countermeasure against a particular security threat.

6

Microsegmentation

Firewall is a common terminology that is widely used in the security field. It is the essential system security component to provide inspections on various networking components. Firewall technologies have been evolved in the past several decades from the simplest dedicated packet filter to today's advanced security appliance that can be easily deployed on any network segments. In addition to guarding against north-south bound 'bad' traffic transported in-and-out of a trusted domain, firewalls have been used to filter east-west traffic within a trusted domain to prevent malicious traffic moving laterally to explore internal vulnerabilities. The granularity level of protected networks within a trust domain can be at the level of a subnetwork, a VLAN, an interface, an application, or a data flow, in which it is usually implemented through a virtual networking approach called *"microsegmentation."*

Microsegmentation is a method of creating secure zones in data centers and cloud deployments to isolate workloads from one another and secure them individually to make network security more granular. The rise of SDN and NFV has paved the way for microsegmentation to be realized in software to ease the deployment and management.

In this chapter, a brief history of firewall and transitions to microsegmentation is firstly described in Section 6.1 and followed with distributed firewall in Section 6.2, microsegmentation system and models are described in Section 6.3, and finally the implementation based on VMWare solutions is presented in Section 6.4.

6.1 From Firewall to Microsegmentation

In Figure 6.1, a brief history of firewalls is presented. The term "firewall" originally referred to a wall intended to confine a fire within a building, i.e., a wall is built to prevent fire from expanding to a larger area. Later uses refer to similar structures, such as the metal sheet separating the engine compartment of a vehicle or aircraft from the passenger compartment. The term was applied in the late 1980s to network technology that emerged when the Internet was fairly new in terms of its global use and connectivity. The predecessors to firewalls for network security were the routers used in the late 1980s.

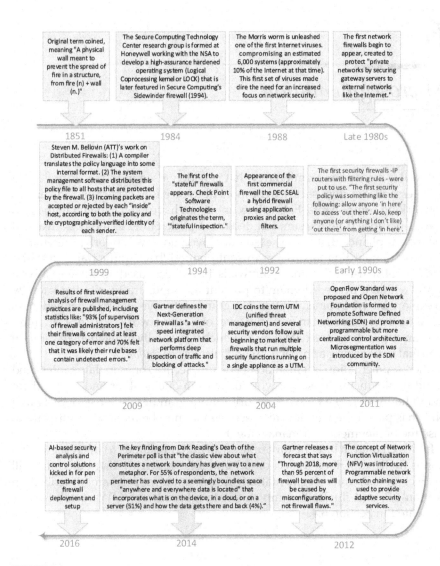

FIGURE 6.1
History of firewall technologies.

The first reported type of network firewall is called a packet filter. Packet filters look at network addresses and ports of packets to determine if they must be allowed, dropped, or rejected. The first paper published on firewall technology was in 1988, when engineers from Digital Equipment Corporation (DEC) developed filter systems known as packet filter firewalls. This fairly basic system is the first generation of what later became a highly involved and technical Internet security feature.

The second-generation firewalls perform the work of their first-generation predecessors but operate up to layer 4 (transport layer) of the OSI model. This is achieved by retaining packets until enough information is available to make a judgment about its state, which is called stateful firewall. A stateful firewall can "remember" a packet session direction, e.g., established TCP connection, to prevent attackers from mimicking a non-returning packet targeting into the protected network system.

Around the late 1990s, firewall inspecting traffic at the application protocol layer started; it is called application-level firewall. The key benefit of application layer firewall is to understand particular applications and protocols such as File Transfer Protocol (FTP), Domain Name System (DNS), or Hypertext Transfer Protocol (HTTP). This is useful as it is able to detect if an unwanted application or service is attempting to bypass the firewall using a protocol on an allowed port, or detect if a protocol is being abused in any harmful way. Time forwards to 2011 and 2012, when SDN and NFV came into the picture, where firewall technologies had faced a huge challenge as well as opportunities. Microsegmentation was brought up in the SDN community to provide fine-grained level of packet inspection at any network segment and push the inspection into a more refined level. As of 2012, the so-called Next-Generation Firewall (NGFW) is nothing more than the "wider" or "deeper" inspection at application stack. For example, the existing deep packet inspection functionality of modern firewalls can be extended to include Intrusion prevention systems (IPS).

Microsegmentation is a method of creating secure zones in data centers and cloud deployments to isolate workloads from one another and secure them individually. It aims at making network security more granular. Network segmentation is not a new concept. Companies have relied on firewalls, virtual local area networks (VLAN) and access control lists (ACL) for network segmentation for years. With microsegmentation, policies are applied to individual workloads for greater attack resistance. The rise of SDN and NFV has paved the way for microsegmentation. Relying on SDN and NFV in software, in a layer that is decoupled from the underlying hardware, this makes segmentation much easier to develop and deploy. Microsegmentation is mainly designed to address two security issues:

- Identify network traffic above layer-4 (transport layer), i.e., user, application, etc.
- Control network traffic in a policy driven manner.

When TCP/IP and Ethernet was gaining a foothold as being the networking platform of choice over token ring and others, it was easy to map OSI layers to devices. For example, hubs are on layer 1, switches are on layer 2, routers are on layer 3, and firewalls are on layer 4. If we want to divide a network into two IP segments, we can use a router, and if we want to segment the network by ports, we then use a firewall to do the work. There are many shortcomings to only using IP and port as a way to segment the network for security purposes. Virtual LAN (or VLAN) is the primary solution today to isolate layer-2 network traffic with the main goal to reduce broadcast traffic. Access Control Lists (ACLs) can help control traffic, but most enterprises have decades old ACLs that are costly to manage. Attackers are able to move around from endpoint to workstation with impunity, which call for more attention to control the east/west traffic flow to effectively secure the network.

There are two main problems with the classic VLAN approach. The first is that a single bad network card sending faulty frames can disrupt the entire network. If you are lucky, this will only affect a single VLAN. If, however, the Network Interface Card (NIC) and switch port are configured to allow trunk access, one bad NIC can affect all VLANs.

The microsegmentation approach is to make use of modern VLAN protocols that are capable not of the classic 4095 VLANs, but instead the 16M VLANs that modern protocols like Shortest Path Bridging (SPB) and VXLAN are capable of. Instead of great big, flat layer 2 networks, the idea is to break everything up such that each virtual network only contains those devices which absolutely must talk to one another and rely on routers to bridge the gaps. Microsegmentation is a promising approach because network administrators have finally started to accept that virtualized routers are useful and usable. For over a decade their use has been anathema, considered by many to be evidence of professional malpractice. As that attitude has changed, so too has both network and software design. Instead of needing expensive, powerful centralized routers to bridge virtual and/or physical networks, virtual machines (VMs) configured as routers can serve the same purpose. Virtualization- and networking-aware management software (VMware's NSX being the canonical example) can dramatically increase security by reducing the number of systems that need to be part of the same virtual network.

What is the 'Micro' in Microsegmentation? Let's say that we have a service that consists of a number of VMs. There is a load balancer, a database, a virtual file server and a bunch of web servers. It is reasonable that these VMs be able to communicate with one another, but there is no good reason for them to communicate with anything else. In a pre-microsegmentation network environment, the VMs that make up this service would likely be part of a virtual or physical network, with hundreds or even thousands of other workloads that were all in the same "zone." Being web-facing, they would probably be part of the "DMZ" zone, and multiple services would be separated from each other through *subnetting*.

Using subnets to isolate workloads is very weak security. It is trivial for an attacker to modify a compromised workload to attempt to access different subnets. It is much harder to get past a properly configured router implementing virtual networks. When VLANs are properly implemented, switch ports – whether virtual or physical – do not allow workloads the opportunity to access arbitrary VLANs. Even in cases (such as virtual routers) where it might make sense to have guest-initiated VLANs, switches are generally configured to only pass packets from VLANs that guest actually needs to access.

This means microsegmentation can be configured such that a given service cannot possibly access other virtual networks except by going through the virtual router; nor can workloads on those other networks access the service you are securing, except through the virtual router. The virtual router, in turn, would only be granted access to the virtual networks that VMs for which it is responsible need to communicate with. Furthermore, with microsegmentation this service would simply be isolated in their own virtual network. If they had a need to talk with another VM, even if it was on the same host, they would go through that host's router. This has numerous security advantages.

In order to transit from traditional firewall approaches to a microsegmentation approach, the **least privilege principle** needs to be implemented:

> *A white-list packet filtering approach needs to be enforced to allow known and legitimated network traffic passing among nodes within a trusted domain.*

Microsegmentation is only possible because management software exists which can relieve the configuration burden. Humans are not good at keeping more than a few hundred interconnections straight in their mind, and when we start isolating each individual service in a large enterprise, we are potentially creating millions of segments. Moreover, microsegmentation is not just about limiting which workloads talk to one another. It is about the automation of network configuration, network service provisioning, and security policy enforcement.

Early development of microsegmentation only limits the impact of any given compromise within a network segment. Simple packet filtering approaches can be established between segment boundaries. It can also be combined with tools that profile services to learn what they communicate with, and then both dynamically configure least-privilege access and exclude potential breaches. Thus, how to maintain connection states is an implementation challenge for microsegmentation considering the large-scale network and system nodes in a datacenter environment. Intelligently providing microsegmentation at a proper granularity is one of the challenging research topics to be addressed.

Other network services beyond just routers are being virtualized. Intrusion detection systems, honeypots and various flavors of automated incident response are all part of the enterprise IT security toolkit today. While these are usable in a classical networking environment, they may also play an important security role in the kinds of highly automated and orchestrated environments that make use of microsegmentation.

6.2 Distributed Firewalls

A firewall is a collection of components interposed between two networks that filters traffic between them according to some security policy. Conventional firewalls depend on the topology restriction of the networks. At the controlled entry point, the firewall divides the networks into two parts, internal and external networks. Since the firewall cannot filter the traffic it does not see, it assumes that all the hosts on the internal networks, e.g., a corporation's network, are trusted and all the hosts on the other side (external), e.g., Internet, are untrusted.

6.2.1 Issues of Conventional Firewalls

Figure 6.2 shows conventional firewalls that serve as the sentry between trusted and untrusted networks. They rely on restricted topology and controlled network entry points to enforce traffic filtering. The key assumption of this model is that everyone on one side of the entry point of the firewall is to be trusted, hence they are protected, and that anyone on the other side is, at least potentially, an enemy. This model works quite well when networks comply with the restricted topology. But with the expansion of network connectivity, such as extranets, high speed lines, multiple entry points, and telecommuting, this model faces great challenges:

- Firewalls do not protect networks from internal attacks. Since everyone on the internal networks is trusted and the traffic within these trusted networks is not seen by the firewall, a conventional firewall cannot filter internal traffics, hence it cannot protect systems from internal threats. For traditional firewalls, the only way to work around this is to deploy multiple firewalls within the internal networks, i.e., divide the network into many smaller networks, and protect them from each other. Since different policies have to be applied on these firewalls, both the load and complexity of administration increase.

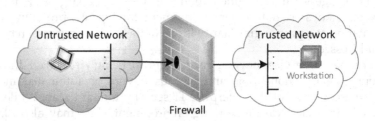

FIGURE 6.2
Conventional firewall: trusted and untrusted domains.

- The vastly expanded Internet connectivity makes this model obsolete. The extranets and the telecommuters from outside are allowed to reach all or part of internal networks. Meanwhile, the telecommuters' computers that use the Internet for connectivity need protection when encrypted tunnels are not in place, especially as cable modems and DSL become more available and affordable. Currently, most such telecommuters connect to organizations' internal networks through VPN tunnels. If they also use the same VPN tunnel for generic Internet browsing purposes, it is not only inefficient (causes the "triangle routing"), but may also violate the organization's guidelines. If they do not use the VPN channel, they either open a security hole or add the workload of maintaining numerous personal firewalls that are at different locations.

- End-to-end encryption is another threat to the firewall. There are so many external Web proxies, such as www.anonymizer.com, out there on the Internet. Users can easily set up an end-to-end encryption tunnel between their desktop within the organization's internal network to a machine outside. Since the firewall does not have the key to look into the encrypted package, it cannot filter properly according to the security policy. By doing so, insider users can bypass the destination restriction and hide the traffic. If this channel is controlled by malicious hackers, it is almost impossible to detect, because all the packages are encrypted!

- Some protocols are not easily handled by a firewall. Because the firewall lacks certain knowledge, protocols like voice-over-IP need application level proxies to manage through the firewall.

- A firewall is the single-entry point. This is the place traditional firewalls enforce their policy and filter the traffic. It is also a single point of failure. If the firewall goes down for any reason, the entire internal networks are isolated from the outside world. Although high availability options, such as hot standby firewall configurations exist, they are usually cost prohibitive.

- Firewalls tend to become network bottlenecks. Due to the increasing speed of networks, the amount of data passing through, and the complexity of protocols firewalls must support (such as IPSec), they are more likely to be the congestion points of networks.

- Unauthorized entry points bypass the firewall security. It has become trivial for anyone to establish a new, unauthorized entry point to the network without the administrator's knowledge or consent. Various forms of tunnels, wireless, and dial-up access methods allow individuals to establish backdoor access that bypasses all the security mechanisms provided by traditional firewalls. While firewalls are in general not intended to guard against misbehavior by insiders, there is a tension between internal needs for more connectivity and the difficulty of satisfying such needs with centralized firewalls. The situation becomes so critical when modern cloud and datacenter networking environment demands tremendous east-west traffic among cloud users.

6.2.2 Introduction of Distributed Firewalls

In order to solve problems of conventional firewalls while still retaining their advantages, Steven Bellovin, an AT&T researcher, proposed a "distributed firewall" [41]. A multitude of host-resident firewalls when centrally configured and managed makes up a distributed firewall. In this architecture, the security policy is still defined centrally, but the enforcement of the policy takes place at each endpoint (hosts, routers, etc.). The centralized policy defines what connectivity is permitted or denied. Then this policy is distributed to all endpoints, where it is enforced. Three components are needed for distributed firewalls:

1. a security policy language;
2. a policy distribution scheme; and
3. an authentication and encryption mechanism, such as IPSec.

The security policy language describes what connections are permitted or prohibited. It should support credentials and different types of applications. After policy is compiled, it is shipped to endpoints. The *policy distribution scheme* should guarantee the integrity of the policy during transfer. This policy is consulted before processing the incoming or outgoing messages. The distribution of the policy can be different and varies with the implementation. It can be either directly pushed to end systems, or pulled when necessary, or it may even be provided to the users in the form of credentials that they use when they try to communicate with the hosts.

How the inside hosts are identified is very important. Conventional firewalls rely on topology. The hosts are identified by their IP addresses and network interfaces on the firewalls they are attached to, such as "inside," "outside," and "DMZ." This kind of structure is quite weak. Anyone with physical access to the internal network and has an internal IP address will be fully trusted, plus IP address-spoofing is not difficult at all. Since all the hosts on the inside are trusted equally, if any of these machines are subverted, they can be used to launch attacks to other hosts, especially to trusted hosts for protocols like rlogin.

It is possible that distributed firewalls use IP addresses for host identification. But a secure mechanism is more desirable. It is preferred to use certificate to identify hosts. IPSec provides cryptographic certificates. These certificates can be very reliable and unique identifiers. Unlike IP address, which can be easily spoofed, the digital certificate is much more secure and the ownership of a certificate is not easily forged. Furthermore, they are also independent of topology. Policy is distributed according to these certificates. If a machine is granted certain privileges based on its certificate, those privileges can be applied regardless of where the machine is physically located.

In this case, all machines have the same rules. They will apply the rules to the traffic. Since they have better knowledge of the connection (such as the state and the encryption keys, etc.), they will make better judgment according to the policy. With a distributed firewall, the spoofing is not possible either, because each host's identity is cryptographically assured. In summary, the distributed firewalls provide the following benefits:

- *Topology Independence*: The most important advantage for distributed firewalls is that they can protect hosts that are not within a topology boundary. The telecommuters who use the Internet both generically and to tunnel in to a corporate network are better protected now. With distributed firewalls, the machines are protected all the time, regardless of whether the tunnel is set up or not. No more triangle routing is needed.

- *Protection from Internal Attacks*: After distributed firewalls abandon the topology restriction, hosts are no longer vulnerable to internal attacks. To the host, there is no more difference between "internal" and "external" networks. After a machine boot up, the policy is enforced on it for any inbound and outbound traffic. In addition, hosts can be identified by their encrypted certificates, this eliminates the chance of identity spoofing.

- *Elimination of the Single Point of Failure*: A traditional firewall needs a single entry point to enforce policy. It not only creates the single point of failure but also limits the entire network's performance to the speed of the firewall. Multiple firewalls are introduced to work in parallel to overcome these problems; in many cases though, that redundancy is purchased only at the expense of an elaborate (and possibly insecure) firewall-to-firewall protocol. With the deployment of distributed firewalls, these problems are totally eliminated. The performance, reliability, and availability no longer depend on one, or in some cases a group of, machines.

- *Hosts Make Better Decisions*: Conventional firewalls do not have enough knowledge in terms of what a host intends. One example is end-to-end encrypted traffic, which can easily bypass the rules on conventional firewalls since the firewalls do not have the necessary key. Also, many firewalls are configured that they will pass the TCP packets from the outside world with the "ACK" bit set, because they think these packets are the replies to the internal hosts who initialize the conversation. Sadly enough, it is not always true and the spoofed ACK packets can be used as part of "stealth scanning."

Similarly, traditional firewalls cannot handle UDP packets properly because they cannot tell if these packets are replies to outbound queries (and hence legal) or they are incoming attacks. In contrast the host that initializes the conversation or sends out the queries knows exactly what packets it is expecting, what it is not, since it has enough knowledge to determine whether an incoming TCP or UDP packets are legitimate and it has the necessary key in the case of end-to-end encryption. The same thing is true for the protocols like FTP.

In addition to the advantages, distributed firewalls have their limitations too. Intrusion detection is harder to achieve on distributed firewalls. Modern firewalls can detect the attempted intrusions. In a distributed firewall, it is not a problem for a host to detect the intrusions; however, the collection of data and aggregate data for intrusion detection purposes is more problematic, especially at times of poor connectivity to the central site, or when either site (host or central site) is under attacks such as DoS.

6.2.3 Implementation of Distributed Firewalls

A project supported by DARPA was fulfilled at the University of Pennsylvania in 2000 [132]. In this project, a prototype of distributed firewall was constructed. OpenBSD was chosen as the operating system, because it was an attractive platform for developing security applications with well-integrated security features and libraries. Keynote [48] was chosen as the security policy language. It was also used to send credentials over an untrusted network. IPSec was used for traffic protection and user/host authentication. This was a concept-proven implementation. Other improvements, such as moving the policy daemon to the kernel and adding IP filters, etc., needed to be done on this prototype firewall.

In a datacenter networking environment, a distributed firewall is an essential component of the overall secure system. It is a network layer, 5-tuple (protocol, source and destination port numbers, source and destination IP addresses), stateful, multi-tenant firewall. As shown in Figure 6.3, when deployed and offered as a service by the service provider, tenant administrators can install and configure firewall policies to help protect their virtual networks from unwanted traffic originating from Internet (i.e., north-south traffic) and intranet networks (i.e., east-west traffic).

The service provider administrator or the tenant administrator can manage the Datacenter Firewall policies via the network controller and the northbound APIs. The Datacenter Firewall offers the following advantages for cloud service providers:

- A highly scalable, manageable, and diagnosable software-based firewall solution that can be offered to tenants.

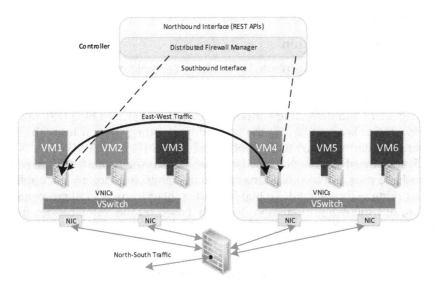

FIGURE 6.3
Distributed firewalls.

- Freedom to move tenant virtual machines to different compute hosts without breaking tenant firewall policies.
  - Deployed as a vSwitch port host agent firewall.
  - Tenant virtual machines get the policies assigned to their vSwitch host agent firewall.
  - Firewall rules are configured in each vSwitch port, independent of the actual host running the VM.
- Offers protection to tenant virtual machines independent of the tenant guest operating system.

The Datacenter Firewall offers the following advantages for tenants:

- Ability to define firewall rules to help protect Internet facing workloads on virtual networks.
- Ability to define firewall rules to help protect traffic between virtual machines on the same L2 virtual subnet as well as between virtual machines on different L2 virtual subnets.
- Ability to define firewall rules to help protect and isolate network traffic between tenant on premise networks and their virtual networks at the service provider.

6.3 Microsegmentation

A natural extension of distributed firewalls is the implementation of microsegmentation. In that sense, we can view microsegmentation is an extension of traditional distributed firewall with more capability on security policy management and defending endpoints. According to Cisco Blog [73], microsegmentation is about having the possibility of setting up policies with endpoint granularity. As of today, both Cisco's ACI [74] and VMWare's NSX [275] are current leading players in the field of microsegmentation. Their solutions have pros and cons in terms of system architecture design and applied networking enthronement. To highlight the salient features of microsegmentation, our following discussion is mainly based on the terminologies from VMWare's implementation [112].

6.3.1 Design Microsegmentation and Considerations

With the growth of SDN and the evolution of software-defined data center (SDDC) technologies, network administrators, data center operators and security officers are increasingly looking to microsegmentation, for enhanced and more flexible network security. To assist in administration, security enforcement and the management of data collision domains, network segmentation or the division of a network into smaller subsections allows administrators to minimize the access privileges granted to people, applications and servers, and the rationing of access to sensitive or mission-critical information on an "as-needed" basis. In what follows, a few important microsegmentation design considerations are presented.

6.3.1.1 Software-Defined and Programmability

In its traditional form, network segmentation is achieved by creating a set of rules to govern communication paths, then configuring firewalls and VLANs to provide the means to partition the network into smaller zones. For comprehensive security coverage, a network should be split into multiple zones, each with their own security requirements – and there should be a strict policy in place to restrict what is allowed to move from one zone to another. It is a physical process that can soon reach the limits of its effectiveness, once traditional or even next-generation firewalls become overloaded and the burden on IT staff of manually fine-tuning configurations and policies becomes too great, as networks expand beyond a certain size or are simply overwhelmed by changing conditions. Microsegmentation takes network partitioning to the next level, by exploiting virtualization and software-defined networking technologies, and utilizing their programmable interfaces to allow policy-based security to be assigned automatically to the network at a granular level with security assignments possible down to individual workloads.

More attention should be put on north-bound interfaces to allow the software-defined system to understand security policies required from applications to specify appropriate east-west traffic based on correctly applied traffic policies enforced between segments.

6.3.1.2 Fine-Grained Data Flow Control and Policy Management

In a datacenter networking environment, for microsegmentation, hardware-based firewalls are usually not required to enable security to be directly integrated with virtualized workloads. As a result, security policies can be synchronized with virtual machines (VMs), virtual networks, operating systems, or other virtual security assets, with security assignments down to the level of a single workload or network interface. When a network is reconfigured or migrated, VMs and workloads will move together with their associated security policies. By enabling such fine-grained security controls, microsegmentation can drastically reduce the available attack surface that a network presents. It enables more granular control over the traditional "choke points" of a network, and allows for security controls that are customized for each virtual environment.

Should an attack occur, the effective separation of each zone, i.e., a microsegment, into its own secured environment helps limit the spread of incursion and any sideways spread into the rest of the network. Microsegmentation can also simplify and speed up incident responses and enhance forensics in the event of a security breach or other network event. On the downside, microsegmentation can be a complex process requiring detailed design and careful administration. The increased overhead in areas like system monitoring and alerts or identity management may translate into increased financial and staffing costs for the enterprise unless the deployment is properly planned and executed.

6.3.1.3 Applying Network Analytic Models to Understand Data Traffic Pattern

Mapping a network's security requirements down to its lowest level requires a detailed knowledge of its inner workings, a fine-grained view of the network that goes beyond what manual observation can achieve. Visibility must be gained into communication patterns and network traffic flows to, from, and within the enterprise campus, and software analytics should be employed to establish key relationships and traffic patterns (groups of related workloads, critical applications, shared services, etc.).

The policies and security rules to be used under microsegmentation will also be determined by the results of network analytics. Models should be drawn up and assessed to highlight important relationships, and to help spot network elements and workloads that may potentially pose problems. Analytical results will also assist in crafting policy definitions and the orchestration system needed for pushing microsegmentation out to all the infrastructure on the network.

6.3.1.4 Zero Trust Zones

Using Microsegmentation, white-list based security policies, i.e., denial of access, should be the default philosophy, with communications on the network selectively allowed on the basis of the previous analysis. Throughout the microsegmentation deployment, "zero trust zones" should be created, with policies and rules set to allow only access to users, systems, and processes that they essentially need to do their jobs. The ideal solution to complete data center protection is to protect every traffic flow inside the data center with a firewall, allowing only the flows required for applications to function. Achieving this level of protection and granularity with a traditional firewall is operationally unfeasible and cost prohibitive, as it would require traffic to be hair-pinned to a central firewall with individual virtual machines placed on distinct VLANs (i.e., Pools of Security model).

6.3.1.5 Tools for Supporting Legacy Networks

Software-defined networking technologies may facilitate the enhancement of legacy infrastructure and security protections for microsegmentation, but this will depend on a careful selection of Hypervisor and tools for virtualization. This would typically include a single tool or platform for visualizing the interactions occurring between the physical and software-defined layers of the network. Tools should also be integrated and user-friendly for all the personnel involved whether they are assigned to operations, networking, cloud, storage, administration, or security.

6.3.1.6 Leveraging Cloud-Based Resource Management and Support

Microsegmentation does incur significant monitoring and management overhead to achieve features such as automated provisioning and move/change/add for workloads, scale-out performance for firewalling, and distributed enforcement in-kernel and at each virtual interface. Cloud-based technologies can relieve much of the burden of a microsegmentation deployment. Network traffic analytics tools may be employed at the design stage to help trace critical communication paths and inter-relationships, and to reduce potential network security and microsegmentation weaknesses, based on known best practice configurations. Deep insights into network operations may be obtained without the need for investing in on-premises hardware and software, and web-based administration platforms may be used to manage and orchestrate the dispersal of microsegmentation policies across the entire network.

6.3.2 Microsegmentation Defined

Microsegmentation decreases the level of risk and increases the security posture of the modern data center. Microsegmentation utilizes the following capabilities to deliver its outcomes:

- *Distributed stateful firewalling*: Reducing the attack surface within the data center perimeter through distributed stateful firewalling. Using a distributed approach allows for a data plane that scales with the compute infrastructure, allowing protection and visibility on a per application basis. Statefulness allows Application Level Gateways (ALGs) to be applied with per-workload granularity.

- *Topology agnostic segmentation*: Providing application firewall protection regardless of the underlying network topology. Both L2 and L3 topologies are supported, agnostic of the network hardware vendor, with logical network overlays or underlying VLANs.

- *Centralized ubiquitous policy control of distributed services*: Controlling access through a centralized management plane; programmatically creating and provisioning security policy through a RESTful API or integrated cloud management platform (CMP).

- *Granular unit-level controls implemented by high-level policy objects*: Utilizing grouping mechanisms for object-based policy application with granular application-level controls independent of network constructs. Microsegmentation can use dynamic constructs including OS type, VM name, or specific static constructs (e.g., Active Directory groups, logical switches, VMs, port groups IP Sets). This enables a distinct security perimeter for each application without relying on VLANs.

- *Network based isolation*: Supporting logical network overlay-based isolation through network virtualization (i.e., VXLAN), or legacy VLAN constructs. Logical networks provide the additional benefits of being able to span racks or data centers while independent of the underlying network hardware, enabling centralized management of multi-data center security policy with up to 16 million overlay-based segments per fabric.

- *Policy-driven unit-level service insertion and traffic steering*: Enabling integration with third-party introspection solutions for both advanced networking (e.g., L7 firewall, IDS/IPS) and guest capabilities (e.g., agentless anti-virus).

6.3.3 NIST Cybersecurity Recommendations for Protecting Virtualized Workloads

The National Institute of Standards and Technology (NIST) is a US federal technology agency working with industry to develop and apply technology, measurements, and standards. NIST works with standards bodies globally in driving forward the creation of international cybersecurity standards. NIST published Special Publication 800-125B [56], "Secure Virtual Network Configuration for VM Protection" to provide recommendations for securing virtualized workloads. The microsegmentation capabilities should satisfy the security recommendations made by NIST for protecting VM workloads.

Section 4.4 of NIST 800-125B [56] makes four recommendations for protecting VM workloads within modern data center architecture. These recommendations are as follows:

- *VM-FW-R1:* In virtualized environments with VMs running delay-sensitive applications, virtual firewalls should be deployed for traffic flow control instead of physical firewalls, because in the latter case, there is latency involved in routing the virtual network traffic outside the virtualized host and back into the virtual network.

- *VM-FW-R2:* In virtualized environments with VMs running I/O intensive applications, kernel-based virtual firewalls should be deployed instead of subnet-level virtual firewalls, since kernel-based virtual firewalls perform packet processing in the kernel of the Hypervisor at native hardware speeds.

- *VM-FW-R3:* For both subnet-level and kernel-based virtual firewalls, it is preferable if the firewall is integrated with a virtualization management platform rather than being accessible only through a standalone console. The former will enable easier provisioning of uniform firewall rules to multiple firewall instances, thus reducing the chances of configuration errors.

- *VM-FW-R4:* For both subnet-level and kernel-based virtual firewalls, it is preferable that the firewall supports rules using higher-level components or abstractions (e.g., security group) in addition to the basic 5-tuple (source/destination IP address, source/destination ports, protocol).

For example, as of 2017, VMware NSX-based microsegmentation meets the NIST *VM-FW-R1, VM-FW-R2,* and *VM-FW-R3* recommendations. It provides the ability to utilize network virtualization-based overlays for isolation and distributed kernel-based firewalling for segmentation with API-driven centrally managed policy control.

6.4 Case Study: VMware NSX Microsegmentation

As shown in Figure 6.4, the VMware NSX platform includes two firewall components: a centralized firewall service offered by the NSX Edge Services Gateway (ESG) and the Distributed Firewall (DFW). The NSX ESG enables centralized firewalling policy at an L3 boundary and provides layer 3 adjacencies from virtual to physical machines. The DFW is enabled in the Hypervisor kernel as a VIB (vSphere Installation Bundle) package on all VMware vSphere® hosts that are part of a given NSX domain. The DFW is applied to virtual machines on a per-vNIC basis.

Protection against advanced persistent threats that propagate via targeted users and application vulnerabilities requires more than L4 segmentation to maintain an adequate security posture. Securing chosen workloads against advanced threats requires application-level security controls such as application-level intrusion protection or advanced malware protection.

NSX-based microsegmentation approaches enable fine-grained application of service insertion (e.g., IPS services) to be applied to flows between assets that are part of a Payment Card Industry (PCI) zone. In a traditional network environment, traffic steering is an all-or-nothing proposition, requiring all traffic to be steered through additional devices. With microsegmentation, advanced services are granularly applied where they are most effective; as close to the application as possible in a distributed manner while residing in a separate trust zone outside the application's attack surface.

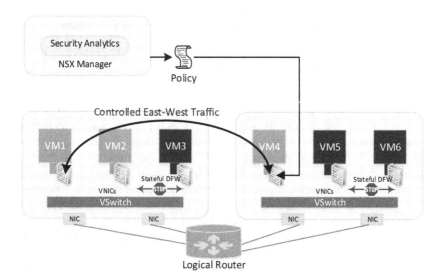

FIGURE 6.4
Distributed segmentation with network overlay isolation.

6.4.1 Isolation

Isolation is the foundation of network security, whether for compliance, containment, or separation of development/test/production environments. Traditionally ACLs, firewall rules, and routing policies were used to establish and enforce isolation and multi-tenancy. With microsegmentation, support for those properties is inherently provided.

Leveraging VXLAN technology, virtual networks (i.e., Logical Switches) are L2 segments isolated from any other virtual networks as well as from the underlying physical infrastructure by default, delivering the security principle of least privilege. Virtual networks are created in isolation and remain isolated unless explicitly connected. No physical subnets, VLANs, ACLs, or firewall rules are required to enable this isolation.

VLANs can still be utilized for L2 network isolation when implementing microsegmentation, with application segmentation provided by the Distributed Firewall. While using VLANs is not the most operationally efficient model of microsegmentation, implementing application segmentation with only the DFW and keeping the existing VLAN segmentation is a common first step in implementing microsegmentation in brownfield environments.

6.4.2 Segmentation

A virtual network can support a multi-tier network environment. This allows for either multiple L2 segments with L3 isolation (Figure 6.5(a)) or a single-tier network environment where workloads are all connected to a single L2 segment using Distributed Firewall rules (Figure 6.5(b)). Both scenarios achieve the same goal of microsegmenting the virtual network to offer workload-to-workload traffic protection, also referred to as east-west protection.

6.4.3 Security Service Function Chaining

Modern data center architectures decouple network and compute services from their traditional physical appliances. Previously, data center operation required traffic to be steered through appliances for services such as firewall, intrusion detection and prevention, and load balancing. As infrastructure services transition from physical appliances to software functions, it becomes possible to deploy these services with greater granularity by directly inserting them into a specific forwarding path. The combination of multiple functions in this manner is referred to as a service chain or service graph, in which two service chaining examples are shown in Figure 6.6.

Once infrastructure services are defined and instantiated in software, they can be created, configured, inserted, and deleted dynamically between any two endpoints in the infrastructure. This allows the deployment and configuration of these services to be automated and orchestrated as part of a Software-Defined Data Center (SDDC).

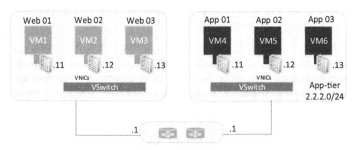

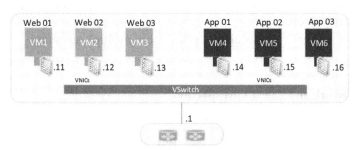

FIGURE 6.5
Segmentation at L3 or using distributed firewalls.

Microsegmentation allows for application-centric, topology-agnostic segmentation. Service insertion in this context permits granular security policies to be driven at the unit or application level rather than at the network or subnet level. This enables the creation and management of functional groupings of workloads and applications within the data center, regardless of the underlying physical network topology.

L2-L4 stateful distributed firewalling features should be provided to deliver segmentation within virtual networks. In some environments, there is a

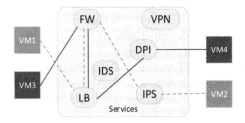

FIGURE 6.6
Two distinct security service chains utilizing different functions.

requirement for more advanced network security capabilities. In these instances, organizations can distribute, enable, and enforce advanced network security services in a virtualized network environment. Network services can be configured into the vNIC context to form a logical pipeline of services applied to virtual network traffic. Network services can be inserted into this logical pipeline, allowing physical or virtual services to be equivalently consumed.

Every security team uses a unique combination of network security products to address specific environmental needs, where network security teams are often challenged to coordinate network security services from multiple vendors. Another powerful benefit of the centralized NSX approach is its ability to build policies that leverage service insertion, chaining, and steering to drive service execution in the logical services pipeline. This functionality is based on the result of other services, making it possible to coordinate otherwise unrelated network security services from multiple vendors.

6.4.4 Network and Guest Introspection

In the NSX infrastructure, there are two families of infrastructure services that can be inserted into an existing topology: *network services* and *guest services*. When deploying a network service, flows are dynamically steered through a series of software functions. For this reason, network services traffic may be referred to as data in motion. Network functions inspect and potentially act on the information stream based on its network attributes. These attributes could include the traffic source, destination, protocol, port information, or a combination of parameters. Typical examples of network services include firewall, IDS/ IPS, and load balancing services.

Guest services act on the endpoints, or compute constructs, in the data center infrastructure. These functions are concerned with data at rest, primarily focusing on compute and storage attributes. Agentless anti-virus, event logging, data security, and file integrity monitoring are examples of guest services.

Service insertion methodologies traditionally rely on network traffic steering to a set of software functions via the physical or logical network control plane. This approach requires an increasing amount of element management and control plane steering as the number of software services scales over time. The NSX microsegmentation-based approach involves steering specified traffic via the NetX (Network Extensibility) framework through one or more Service Virtual Machines (SVMs). These SVMs do not receive network traffic through the typical network stack; they instead are passed traffic directly via a messaging channel in the Hypervisor layer. Network traffic designated for redirection to a third-party service is defined in a policy-driven manner utilizing a feature called Service Composer. The traffic steering rules are based on defined security policies.

This framework for traffic redirection is known as VMware Network Extensibility or NetX. The NetX program features a variety of technology partners

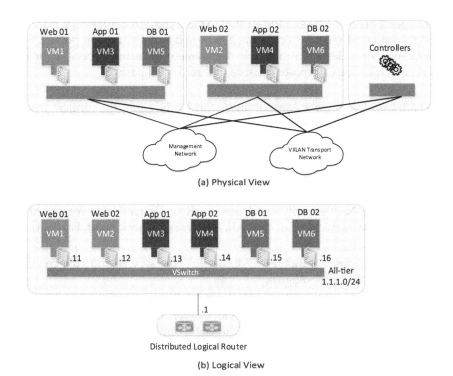

FIGURE 6.7
Security abstracted from physical topology.

across the application delivery, security, operations, and inter-domain feature sets, with additional partners constantly joining the ecosystem.

6.4.5 Security Service Abstraction

Network security has traditionally been tied to network constructs. Security administrators had to have a deep understanding of network addressing, application ports, protocols, network hardware, workload location, and topology to create security policy. As shown in Figure 6.7, using network virtualization can abstract application workload communication from the physical network hardware and topology; this frees network security from physical constraints, enabling policy based on user, application, and business context.

6.4.5.1 Service Composer

Microsegmentation enables the deployment of security services independent of the underlying topology. Traditional (e.g., firewall) or advanced (e.g., agentless Anti-Virus (AV), L7 firewall, IPS, and traffic monitoring) services can be deployed independent of the underlying physical or logical networking topologies. This enables a significant shift in planning and deploying services in the data center. Services no longer need to be tied to networking topology. Service Composer is designed to enable deployment of security services for the data center. Service Composer consists of three broad parts:

- *Intelligent Grouping*: decoupling workloads from the underlying topology via creation of security groups.
- *Service Registration and Deployment*: enabling third-party vendor registration and technology deployment throughout the data center.
- *Security Policies*: security policies allow flexibility in applying specific security rules to distinct workloads. Policies include both governance of built-in NSX security services as well as redirection to registered third-party services (e.g., Palo Alto Networks, Check Point, Fortinet).

There are various advantages in decoupling the service and rule creation from the underlying topologies:

- *Distribution of Services*: The services layer allows distribution and embedding of services across the data center, enabling workload mobility without bottlenecks or hairpinning of traffic. In this model, granular traffic inspection is done for all workloads wherever they reside in the data center.
- *Policies are Workload-Centric*: Policies are natively workload-centric rather than requiring translation between multiple contexts - from workloads to virtual machines to basic networking topology/IP address constructs. Policies can be configured to define groups of database workloads that will be allowed specific operations without explicitly calling out networking-centric language (e.g., IP subnets, MACs, ports).
- *Truly Agile and Adaptive Security Controls*: Workloads are freed from design constraints based on the underlying physical networking topologies. Logical networking topologies can be created at scale, on demand, and provisioned with security controls that are independent of these topologies. When workloads migrate, security controls and policies migrate with them. New workloads do not require the recreation of security polices; these policies are automatically applied. When workloads are removed, their security policies are removed with them.
- *Service Chaining is Policy-based and Vendor Independent*: Service chaining is based on a policy across various security controls. This has evolved from manually hardwiring various security controls from multiple vendors in the underlying network. Policies can be created, modified, and deleted based on the individual requirements.

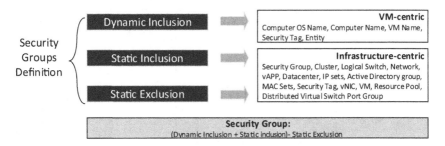

FIGURE 6.8
Security grouping.

6.4.5.2 Grouping

Grouping mechanisms can be either static or dynamic in nature, and a group may be any combination of objects. Grouping criteria can include any combination of objects, VM Properties, or Identity Manager objects (e.g., Active Directory Group). A security group is based on all static and dynamic criteria along with static exclusion criteria defined by a user. Figure 6.8 details the security group construct and valid group objects.

Static grouping mechanisms are based on a collection of virtual machines that conform to set criteria. Objects define core networking and management components in the system. Combining different objects for the grouping criteria results in the creation of the *AND* expression. Dynamic grouping mechanisms are more flexible; they allow expressions that define the virtual machines for a group. The core difference from static grouping mechanisms is the ability to define AND/OR as well as ANY/ALL criteria for the grouping. Evaluation of VMs that are part of a group is also different. In a static grouping mechanism, the criteria instruct on which objects to look for in terms of change. In dynamic grouping, each change is evaluated by the controllers in the data center environment to determine which groups are affected.

As dynamic grouping has a greater impact than static grouping, evaluation criteria should look at zones or applications and be mapped into sections to avoid the global evaluation of dynamic objects. When updates or evaluations are performed, only the limited set of objects belonging to a specific section are updated and propagated to specific hosts and vNICs. This ensures rule changes only require publishing of a section rather than the entire rule set. This method has a critical impact on performance of controllers and the propagation time of policy to hosts.

6.4.5.3 Intelligent Grouping

Change Firewall configuration is not a trivial task. In the following example, two virtual machines are added to a logical switch. In a traditional firewall scenario, IP addresses or subnets must be added or updated to provide

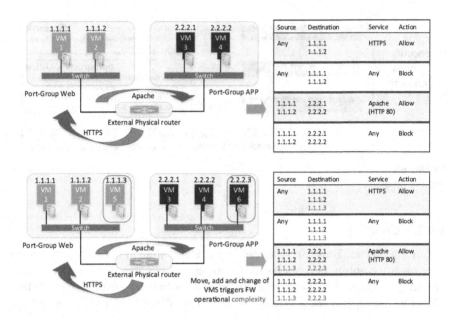

FIGURE 6.9
Traditional firewall rule management overhead.

adequate security to the workload. This implies that the addition of firewall rules is a prerequisite of provisioning a workload, i.e., for every Firewall rule entry related to the added VM will need to be added. When the network system is large, manually changing these rules is prone to errors. Moreover, it is not easy to check potential rule conflicts with manually added rules.

Figure 6.9 diagrams this process. With Security Grouping (SG), a security or network administrator can define that any workload on this VM will be assigned similar security controls. With intelligent grouping based on the NSX Logical Switch, a rule can be defined that does not change with every VM provisioned. The security controls adapt to an expanding data center as shown in Figure 6.10. By using SG, the rules do not need to be changed

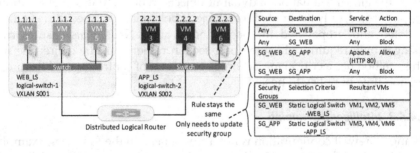

FIGURE 6.10
Adaptive security with Security Group.

when a newly configured VM is added into the system. Instead, changes only need to be done by configuring the security groups. In this way, the configuration overhead can be significantly reduced.

6.4.5.4 Security Tag

Security tags can be applied to any VM. This allows for classification of virtual machines in any desirable form. As shown in Figure 6.11, two tags 'FIN-TAG-WEB' and 'HR-TAG-WEB' are used for finance department and HR department, and they belong to security groups SG-FIN-WEB and SG-HR-WEB, respectively. Furthermore, tags 'FIN-WEB-01' and 'FIN-WEB-02' are used for identify web traffic to/from two individual VMs in the SG-FIN-WEB group.

Some of the most common forms of classification for using security tags are:

- Security state (e.g., vulnerability identified);
- Classification by department;
- Data-type classification (e.g., PCI Data);
- Type of environment (e.g., production, developments);
- VM geo-location.

Security tags are intended for providing more contextual information about the workload, allowing for better overall security. In addition to users creating and defining their own tags, third-party security vendors can use the same

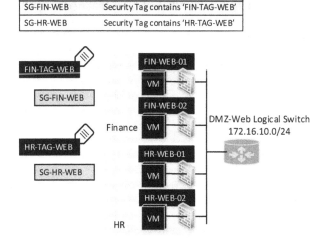

FIGURE 6.11
Tag used as base isolation method.

tags for advance workload actions. Examples of vendor tags include trigger on malware found, suspicious workload activity, and Common Vulnerabilities and Exposures (CVE) score compliance. This functionality allows for context sharing across different third-party vendors.

Summary

In this chapter, we illustrated the transition from traditional firewall to modern microsegmentation approaches by using software-defined and virtual networking approaches. The enabling technique of microsegmentation is due to the significantly improved capacity of networking devices and servers' hardware supporting both virtualized networks and software security appliances. However, several challenges still exist that make the deployment of microsegmentation solutions still in its early stage.

The first challenge is to maintain states of network traffic flows, which is a critical service to support security functions such as stateful firewall, intrusion detection and prevention, etc. However, it is a non-trivial task to maintain all network traffic flow states, especially, when maintaining states at each individual interface (both physical and virtual interfaces) level. Moreover, it requires to design a distributed state monitoring system for decentralized state management that can be used to support realtime traffic analysis at each network segment.

The second challenge is to build an effective security and traffic analysis model. Traditional centralized data processing model cannot satisfy realtime security decisions for security services such as DFW, IPS, DPI, etc. Both of these challenges are still open research problems that need significant research efforts to overcome.

7

Moving Target Defense

Static nature of cloud systems is useful from a service provisioning aspect. The cloud service providers want system configuration to remain unchanged once an application has been deployed. This makes the cloud system soft target for the attackers since they can spend the time to perform reconnaissance on the system and craft the necessary attacks based on the information gathered by cloud system exploration. The static and homogeneous nature of the cloud system, although it makes administration easy, increases the chance of a system being compromised.

MTD has been used as a defensive technique in many fields such as one-on-one air combats, Go game, chess, etc. The goal is to deceive the attacker. MTD allows the administrator to change the static nature of cloud resources. The cloud system information that is accessible to the attacker, such as open ports, Operating System (OS) information, and software version information together constitute the *Attack Surface*. By introducing MTD, the static nature of a cloud system can be changed to dynamic. The homogeneous attack surface becomes asymmetric and heterogeneous. The constantly changing attack surface reduces the probability of successful exploits by the attacker.

In this chapter, we introduce MTD-based proactive security. The introduction of cyber kill chain and how MTD can help in disruption of attack propagation at various stages of attack have been discussed in Section 7.1. The classification of different types of MTD mechanisms, along with illustrative examples, has been discussed in Section 7.2. We consider some examples of SDN-based MTD frameworks that utilize Service Randomization, OS hiding and other obfuscation techniques using SDN-based command and control have been discussed in Section 7.3. Section 7.4 considers MTD as a game between attacker and defender and discusses existing approaches that leverage game theoretic detection and defense mechanisms to deal with security attacks. The evaluation of the effectiveness of different MTD frameworks has been provided in Section 7.5.

7.1 Introduction

Today's security teams consider monitoring the network, detection of incidents, prevention and remediation of security breaches as the most important

FIGURE 7.1
Phases of Cyber Kill Chain.

tasks. The assumption is that infrastructure will remain static over time and security teams try to defend it using state-of-the-art security analysis tools. The threat vectors and adversaries are however constantly evolving. The polymorphic and dynamic adversaries enjoy the advantage of an unchanging attack surface, which they can study at their leisure.

It is a well-known fact that a moving target is harder to hit compared to a static target. When the intruders are able to successfully breach a certain endpoint in the cloud system, they can use that as a way to further exploit critical services, and gain complete command and control of the network.

The cyber kill chain in Figure 7.1 comprises courses of action an attacker is likely to take while exploiting a system. MTD can disrupt the attack propagation by increasing the cost of attack for the attacker in the early stages of the chain. For instance OS rotation [266] will render knowledge gained by an attack by probing the network (*Reconnaissance*) redundant. The payload that attackers might have crafted (*Weaponization*) against a particular OS version, e.g., Ubuntu 14.04, to exploit a system vulnerability will not work on an alternate version of OS - Ubuntu 16.04. There are several factors such as timing, cost, valid configurations that need to be considered while using MTD as discussed in [297]. The MTD techniques should minimize the threat level in the network and at the same time limit service disruption for legitimate users.

7.2 MTD Classification

Moving Target Defense (MTD) can be classified based on two approaches, i.e., classification based on security modeling and classification based on placement in the IP protocol stack. We discuss both these approaches in this section.

7.2.1 Security Modeling-based MTD

MTD techniques based on security modeling can be classified into (i) Shuffle, (ii) Diversification and (iii) Redundancy as discussed by Hong et al. [118].

7.2.1.1 Shuffle

A shuffle technique involves rearrangement of resources at various layers. Some examples of a shuffle technique include migration of VM from one physical server to another, application migration, instruction set randomization (ISR), etc.

Figure 7.2 shows two servers under attack. The network administrator can take the shuffle action as shown above to reconfigure network topology in such a way that both attackers are connected to *Server 2* in the new configuration. The legitimate users - *Users 1,2,3* - can use *Server 1*, and attackers can be quarantined to a specific zone of the network. This will slow down the attack propagation and increase the cost of attack for the attackers. *Server 2* can also act as honeypot post-network reconfiguration if the administrator wants to study the attacker's behavior.

7.2.1.2 Diversification

Diversification-based MTD techniques modify the network function or software responsible for the functioning of an application or the underlying compiler to diversify the attack surface as discussed in [283]. The accretion,

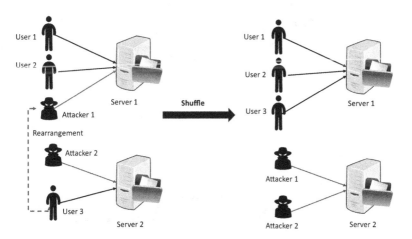

FIGURE 7.2
MTD shuffle.

excision, and replacement of functionality are some examples of software-based diversification techniques.

Security attacks like *Code Reuse Attacks (CRAs)* that use jump oriented programming approach and library code reuse techniques [49] can be eliminated by diversity based MTD techniques [116]. CRAs exploit the fact that all binaries of a given version of software are identical. Attackers can use reverse engineering on one binary to identify regions of code with vulnerability, and then craft a payload that works on all such binaries. However, if the target binaries are different, then the gadget blocks identified from one become useless on the other binary.

Figure 7.3 shows container-based software diversification solution. The paradigm of software delivery has shifted from physical delivery of software via media to application stores such as *Amazon Web Server (AWS)* [67]. Lightweight, executable piece of software - *Container* [191] - can be downloaded and uploaded to container stores like AWS.

The application developer can create and push the application to the container store. The application delivery engine is an MTD diversification-based application randomizer, which produces a different variant of the base software binary represented by different icons in Figure 7.3 on each client request. Thus an attacker who wants to target application needs to reverse-engineer not only the memory gadgets from the host binary, but also the diversification approach used in order to craft an exploit payload against all variants of a software version. This increases the cost of attack for the attacker by a large amount.

The diversification approach can also be applied to OS version and runtime environments, e.g., *Address Space Layout Randomization (ASLR)* [246] and *Instruction Set Randomization (ISR)*.

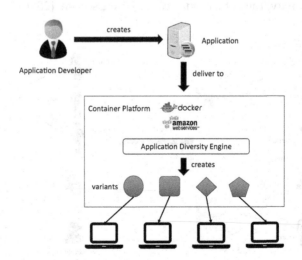

FIGURE 7.3
MTD diversification.

7.2.1.3 Redundancy

Redundancy creates multiple replicas of a network component in order to maintain an optimal level of service in case of network attacks like DDoS attacks. Another benefit of redundancy is that we can create decoys in the network for increasing the discovery time of the actual target. The goal of the attacker in a DDoS attack is saturation of network nodes. With redundancy, the attack goal is hard to achieve for the attacker, since the administrator can load balance traffic to different proxies in case of traffic surge and maintain network reliability.

A layer of proxies between the client and the application accessed by the client as discussed by MOTAG [136] architecture can be used to inhibit the direct attack on the network infrastructure. As shown in Figure 7.4, the client's requests to legitimate services hosted on application server need to pass through an authentication server.

The actual IP address of the application server is concealed from the clients, and a group of proxy nodes acts as a relay between the clients and the server. Once the authentication server verifies the client access, the client can connect to one of the proxy nodes available and access the application server. The proxies only talk to the authentication server so they are resilient to the attacks. Also, the proxy nodes keep changing proxy nodes dynamically, so the attacker is confused about the actual location of the proxy node.

7.2.2 Implementation Layer-based MTD

Another way of classifying MTD techniques is based on the implementation in the protocol stack. The level at which decisions of making a change in attack

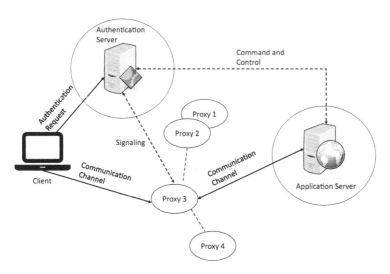

FIGURE 7.4
MTD redundancy.

surface can involve some configuration change inside the host, application binary or the way networking and routing have been configured in the target environment. We classify MTD based on the implementation in protocol stack into categories below (i) Network Level MTD, (ii) Host Level MTD, and (iii) Application Level MTD.

7.2.2.1 Network Level MTD

Network Level MTD requires a change in the network parameters, such as network configuration, network connection obfuscation and Routing Randomization. By making changes to the network configuration, several key issues such as zero-day attacks and, false positives from IDS can be mitigated. Such a countermeasure, however, also affects the service availability of the genuine users. An intelligent system can take operational and security goals of the system into account and utilize intelligent MTD adaptations in order to secure the network as discussed by Zhuang et al. [298].

An overview of the various components involved in Network MTD has been discussed by Green et al. [101]. There are some clients who behave in accordance with the network and security policies. Such clients are referred to as *trusted clients*, as shown in Figure 7.5. A service important from the perspective of the normal functioning of business is often the target of the attackers. The attacker's goal is to disrupt the service or steal key information from the application or database by exploiting vulnerability present on the service. Such services are referred to as *target*. A decoy put in place to confuse the attacker carrying no useful information is known as *sink*. The *mapping*

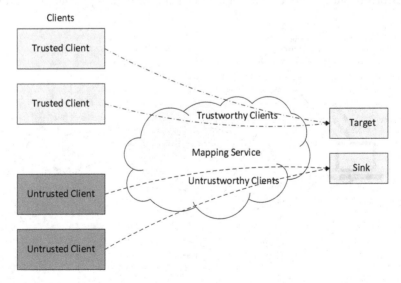

FIGURE 7.5
Overview of components of network MTD system.

system consists of an access control mechanism in order to distinguish the legitimate users from the attackers. Once the clients have been authenticated, they are authorized to access the network services.

The network-based MTD system consists of three key properties:

1. *Moving Property* which requires a continual change of information or configuration by forcing clients to reach the services using mapping system and limit the exposure to target services by shrinking the window of opportunity for the attackers.

2. *Access Control Property* ensures that policies for accessing the network services are enforced properly. Only the authenticated and properly authorized users should be able to use the target services.

3. *Distinguishability Property* to separate the trustworthy clients from untrustworthy users based on network activity. Any classification errors in this property can result in blocking of the trusted users from accessing the services or allow untrustworthy users to exploit the target in the network.

7.2.2.2 Host Level MTD

Host Level MTD requires a change in host resources, OS, and renaming of configurations. Multi-OS Rotation Environment (MORE) [266] is an example of host level MTD. The system consists of two sets of IP addresses, i.e., *Live IP* and *Spare IP*. The live IP address can be accessed by the clients connected to the host, whereas the spare IP address is used when the host is being rotated. The host which is selected during the rotation phase is analyzed for the presence of malware, evidence of intrusion attempts, etc. If the host is found to be compromised, it can be replaced by the clean version of OS or the affected service. The framework is managed by an administrator daemon, which aims to achieve high service availability during the rotation process.

7.2.2.3 Application Level MTD

Application Level MTD involves a change in the application required, source code, memory mapping, and software version. An example of software-based MTD is randomization of application address space. Address Space Layout Randomization (ASLR) is a technique in OS to hide the base address of the software gadgets running on the OS. The ASLR protection forces the attacker to guess the location of the gadget.

However, some parts of the file are not randomized, leaving weak spots in the application for the attacker, which can be used by him to invoke malicious code. Most of the software exploits are based on a technique known as Return-Oriented Programming (ROP). ROP is an exploit technique which allows an attacker to take control of the program flow by smashing the call stack and inserting the malicious instruction sequence in the program. ROP is able to

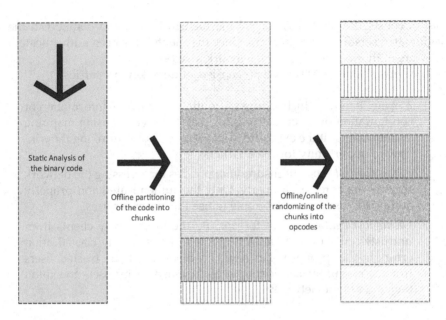

FIGURE 7.6
Anti-ROP application level MTD.

bypass Data Execution Prevention (DEP), a technology used in hardware and software to prevent code injection and executions.

MTD can be used to prevent ROP attacks. The attackers leveraging the ROP attacks assume that the gadgets they are targeting are in absolute address or shifted by some constant bytes. MTD solution can be used to ensure that even if the attacker can detect the byte shift of one gadget, he is not able to guess byte shift for the rest of the gadgets.

The Process Executable (PE)/ELF file for the application binary can be analyzed and the chunks of different memory blocks can be rearranged in order to provide MTD based security for the application as shown in Figure 7.6. The reorganized binary file provides the same functionality as the original application binary. The attacker, in this case, will need to find the address of all the gadgets in order to successfully exploit the application.

7.3 SDN-based MTD

SDN allows centralized command and control of the network. The flexible programmable network solution offered by SDN makes it an ideal candidate for provisioning MTD solutions. For instance, in a cloud network managed by SDN, the SDN controller can be notified by security analysis tools about active threat in the network, and the controller can take preventive methods to deal with the situation such as IP address hopping, reconfiguration of routes for network traffic or changing the host endpoint to delay the attack propagation. In this section, we analyze some techniques that involve the use of SDN.

7.3.1 Network Mapping and Reconnaissance Protection

The first step of the Cyber Kill Chain is the identification of vulnerable software and OS versions. Most scanning tools make use of ICMP, TCP or UDP scans to identify the connectivity and reachability of the potential targets. The replies to the scans can also reveal the firewall configuration details, i.e., what traffic is allowed or denied. The Time to Live (TTL) information can also help in the identification of a number of hops to the attack target [143].

SDN-enabled devices can use MTD adaptations can be used to delay the attack propagation by masquerading the real response and replying back with a fake response to confuse the attacker. As a result, the scanning tool will see random ports open in the target environment. The attacker's workload will be increased because he will have to distinguish the fake reply from the real traffic reply. The SDN-enabled devices can also introduce random delays in TCP handshake request that will disrupt the identification of TCP services and as a result help in the prevention of Distributed Denial of Service (DDoS) attacks. Kampanakis et al. [143] have discussed the cost-benefit analysis of MTD adaptations against network mapping and reconnaissance attacks.

7.3.1.1 Service Version and OS Hiding

The attacker needs to identify the version of OS or vulnerable service in order to mount an attack. For instance, the attacker can send *HTTP GET* request to Apache Web Server, and the response can help in identification of vulnerability associated with a particular version of the Apache software. If the attacker gets a reply *404 Not Found*, he can identify some obfuscation happening at the target software. A careful attacker can thus change the attack vector to exploit the vulnerability at the target.

An SDN-enabled solution can override the actual service version with a bogus version of the Apache Server. Some application proxies leverage this technique to prevent service discovery attempts by a scanning tool.

Another attack method known as *OS Fingerprinting*, where the attacker tries to discover the version of the operating system which is vulnerable. Although modern OS can generate a random response to TCP and UDP requests, the way in which TCP sequence numbers are generated can help an attacker in the identification of OS version.

In an SDN-enabled solution, the OS version can also be obfuscated by the generation of a random response to the probes from a scanning tool. SDN can introduce a layer of true randomization for the transit traffic to the target. The SDN controller can manage a list of OS profiles and send a reply resembling TCP sequence of a bogus OS, thus misguiding the attacker. As shown in

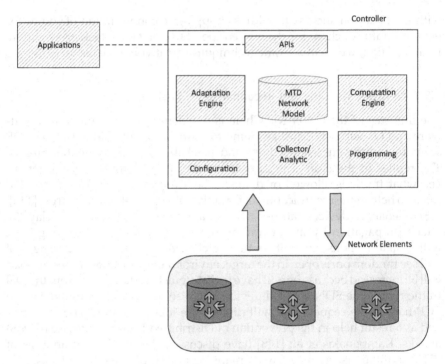

FIGURE 7.7
SDN-based MTD solution.

Figure 7.7, the SDN controller can use computation engine, and collector/ analytic modules to analyze the traffic statistics, and the MTD adaptation engine can generate service version and OS hiding based MTD strategies.

7.3.2 OpenFlow Random Host Mutation

SDN makes use of OpenFlow protocol for control plane traffic. Jafarian et al. [133] proposed OpenFlow enabled MTD architecture as shown in Figure 7.8, can be used to mutate IP address with a high degree of unpredictability while keeping a stable network configuration and minimal operational overhead.

The mutated IP address is transparent to the end host. The actual IP address of the host called real IP (*rIP*) is kept unchanged, but it is linked with a short-lived virtual IP address (*vIP*) at regular interval. The vIP is translated before the host. The translation of rIP-vIP happens at the gateway of the network, and a centralized SDN controller performs the mutation across the network. A Constraint Satisfaction Problem (CSP) is formulated in order to maintain mutation rate and unpredictability constraints. The CSP is solved using Satisfiability Modulo Theories (SMT) solver.

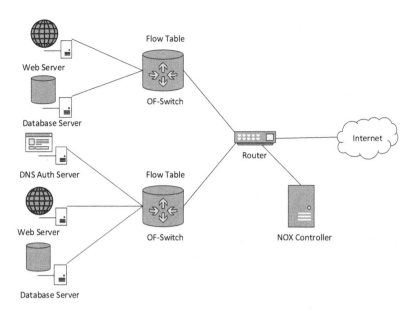

FIGURE 7.8
OpenFlow random host mutation.

Sensitive hosts have a higher mutation rate compared to the regular hosts in this scheme. The OF-RHM is implemented using *mininet* network simulator and *NOX* SDN controller. OF-RHM is able to thwart about 99% of information gathering and external scanning attempts. The framework is also highly effective against worms and zero-day exploits.

7.3.3 Frequency Minimal MTD Using SDN

Frequency minimal MTD [79] approach considers resource availability, QoS and exploits probability for performing MTD in an SDN-enabled cloud infrastructure.

The design goals of the approach consider:

1. What is the optimal frequency at which proactive VM migration should take place, provided wastage of cloud resources is minimal?
2. What should be the preferred location for VM migration which can be selected without impacting the application performance?

As shown in Figure 7.9, the normal clients can access the services hosted in the cloud network via *regular path*, and the *attack path* represents the path along which the attacker tries to exploit the target application.

The VMs periodically share their resource information such as storage, compute capacity with the controller along the *control path*. Depending upon the level of threat, the migration of VM can be proactive or reactive. The real IP

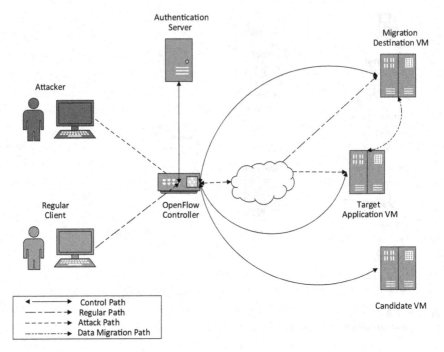

FIGURE 7.9
VM migration based MTD against DoS attack.

address of the VM hosting cloud application is hidden from the clients. The *data path* shows the path along which VM application and its back-end database are migrated. The VM migration is based on factors:

- **VM Capacity:** This parameter considers the capacity of migration target in terms of computing resources available to store files and database of current and future clients.

- **Network Bandwidth:** The lower the network bandwidth between the source and target VM, the slower will be the migration process and the longer will be the exposure period (in case of active attacks) for the VM being migrated. This parameter considers bandwidth between source and target while performing VM migration.

- **VM Reputation:** This is the objective indicator of VM robustness to deter future cyber attacks. It is the history of VM in terms of cyber attacks launched against the VM. This parameter is considered in order to ensure VM's suitability for migration.

This research work estimates the optimal migration frequency and ideal migration location based on the parameters described above. The VM migration mechanism is highly effective in dealing with denial-of-service (DoS) attacks.

7.3.4 SDN-based Scalable MTD in Cloud

SDN-based scalable MTD solution [64] makes use of Attack Graph-based approach to perform the security assessment of a large-scale network. Based on the security state of the cloud network, MTD countermeasures are selected. Figure 7.10 shows system modules and operating layers, which are part of SDN-based MTD framework. The overlay network is responsible for vulnerability analysis, attack graph generation. The physical network consists of OVS, running on top of the physical server. The SDN control interacts with OVS using OpenFlow APIs.

The security analysis of a large-scale cloud network in real time is a challenging problem. Attack Graphs help in identification of possible attack scenarios that can lead to exploitation of vulnerabilities in the cloud network. The attack graphs, however, suffer from scalability issues, beyond a few hundred nodes.

The research work utilizes an approach based on Parallel Hypergraph Partitioning [145] in order to create a scalable attack graph in real time and select MTD countermeasure - VM migration. The MTD strategy considers the available space on the target physical server and the threat level of the destination server before performing the VM migration.

The SDN framework also checks the possible conflicts in the OpenFlow rules after MTD countermeasure deployment. Six types of security conflicts, i.e., (i) Redundancy, (ii) Generalization, (iii) Correlation, (iv) Shadowing, (v) Overlap, and (vi) Imbrication are identified based on overlap in match and action fields of flow rules. The flow rule extraction, conflict detection, and resolution happen at the logical layer in cloud network as shown in Figure 7.10. The conflicted flow rules are identified and corrected in order to provide security compliance in SDN environment post-MTD countermeasure.

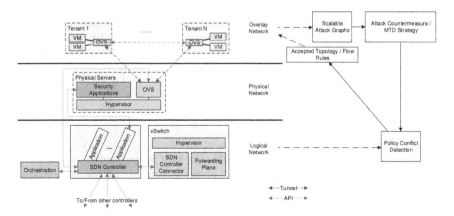

FIGURE 7.10
System modules and operating layers of scalable MTD solution.

7.4 Game Theoretic MTD Models

MTD can be considered as a game between the defender and the attacker. The goal of the attacker is the exploit of critical services in the network with a minimum possible cost of attack. The goal of the defender is to invest the security budget in such a way that the cost of attack is maximized for the attacker. The defensive strategy should stop or slow down the attack propagation. In this section we analyze the game theoretic model for MTD that utilizes various reward metrics such as time, network bandwidth, service latency, etc., to study the MTD techniques.

7.4.1 Game Theoretic Approach to IP Randomization

The framework comprises of real nodes that the attacker wants to target, and the decoy nodes, which have been added to the system by the defender in order to distract and slow down the attacker [66]. A game theoretic model is used to study the strategic interactions between the attacker and the decoy nodes.

The framework consists of two parts. The *first* part evaluates the interaction between the adversary and a single decoy node. The adversary in this step can use the variation in timing from real nodes and decoy nodes in order to identify the actual target nodes. Another way that can be used by the adversary to target-decoy nodes is a study of variation in protocol implementation at real and decoy nodes.

The equilibrium of the game in the first stage is used to formulate a game with multiple real nodes and decoy nodes in the *second* stage. The goal of an adversary in the attack model is to discover the real node's IP address in order to select the appropriate attack against services running on that VM.

The information set of the system can be described by the current number of valid sessions, Y(t) and decoy node fraction scanned by the adversary at time 't', denoted by D(t). The goal of the system administrator is to minimize the cost function composed of an information set by choosing an optimal random variable R:

$$min_R E(D(R)) + \beta Y(R). \tag{7.1}$$

The β denotes delay introduced by migration of a connection to the IP address of the real node from decoy node. The cost of the connection is, therefore, $\beta Y(t)$. The randomization policy is the mapping of the information space (Y(t), D(t)) to {0,1} random variable, where a strategy of not randomizing at the time 't' is denoted by 0 and the strategy of randomizing at the time 't' is denoted by 1.

7.4.2 Game Theoretic Approach to Feedback Driven Multi-Stage MTD

Computer networks suffer from network scanning and packet sniffing tools which are used by a malicious attacker for information gathering. A multi-stage defense mechanism [296] can be formulated based on feedback information structure. The multi-stage defense mechanism manipulates the attack surface in order to provide a proactive defense mechanism against attacks.

The attacker needs to learn the network setup continuously, and change his attack vector accordingly. This is known as attack cost. The defender, on the other hand, needs to consider the reconfiguration cost of shifting the attack surface. Such interactions between attacker and defender can be formulated as a game, in which defender has the objective of minimizing the security risk while maintaining the usability of the system. Thus, the defender has to find the optimal configuration policy in order to achieve the desired objective. The attacker in this game has the objective of exploring the attack surface and inflicting maximum damage on the system.

Consider a system to be partitioned into several layers, $l = 1,2,3,...,N$. The vulnerabilities in each layer are denoted by $V_l := \{v_{l,1}, v_{l,2}, ..., v_{l,n_l}\}$. Each vulnerability is a system weakness that can be exploited by attacker. The vulnerability set V_l is common knowledge for both attacker and defender. Each system configuration can have one or many vulnerabilities. Let the set of feasible system configurations at layer l be depicted by $C_l : \{c_{l,1}, c_{l,2}, ..., c_{l,m_l}\}$. The function π_l maps associate each system configuration with vulnerability set, i.e., $\pi_l : C_i \rightarrow 2^{V_i}$.

Figure 7.11 consists of four layers, $l = 1,2,3,4$ and an attack surface with vulnerability set $V_l = \{v_{l,1}, v_{l,2}, v_{l,3}\}$. There are two possible feasible configurations at layer 1, i.e., $C_1 = \{c_{1,1}, c_{1,2}\}$. In Figure 7.11 the configuration chosen is $c_{1,1}$. The configuration is subject to two vulnerabilities, $\pi_1(c_{1,1}) = \{v_{1,1}, v_{1,2}\}$.

Similarly $\pi_2(c_{2,1}) = \{v_{2,1}, v_{2,2}\}$, $\pi_3(c_{3,1}) = \{v_{3,1}, v_{3,2}\}$, $\pi_1(c_{4,2}) = \{v_{4,3}, v_{4,3}\}$. The configuration in this example $\{c_{1,1}, c_{2,1}, c_{3,1}, c_{4,2}\}$ allows the attacker to launch a successful attack through a sequence of exploits $v_{1,1} \rightarrow v_{2,2} \rightarrow v_{2,3} \rightarrow v_{4,3}$.

If the system configuration remains static, an attacker can launch a multi-stage attack by systematically scanning and exploiting vulnerability corresponding to each configuration. A mixed strategy can be employed by the defender as shown in Figure 7.12.

The defender changes the configuration at *stage 1* to $c_{1,2}$. Thus the overall system configuration changes to $\{c_{1,2}, c_{2,1}, c_{3,1}, c_{4,2}\}$. The attack planned by an attacker based on original network configuration will not succeed in this case. The attack-defense interaction can be formulated as two-player zero-sum game. The defender can randomize the attack surface at each layer to thwart multi-stage attacks.

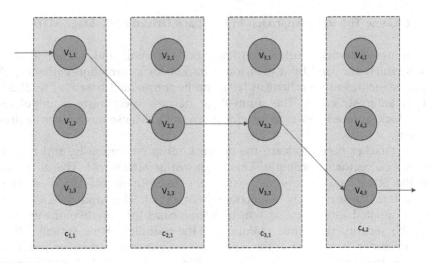

FIGURE 7.11
Static configuration and sequence of attack propagation on a physical system. Solid curve is used to denote existing vulnerabilities.

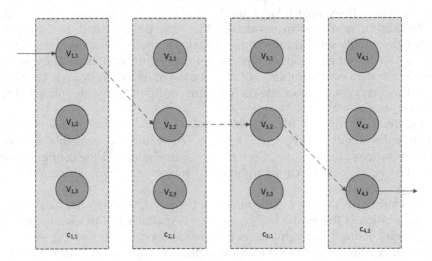

FIGURE 7.12
Randomized configuration of attack surface at layer 1. Dotted curves show vulnerabilities circumvented by current configuration.

7.4.3 Game Theory-based Software Diversity

The over-reliance of a certain version of the software and an operating system can lead to security compromise propagation over a large domain of networks or cloud systems. The diversity as a solution to software monoculture problem has been studied by Neti et al. [203]. The hosts in the network and

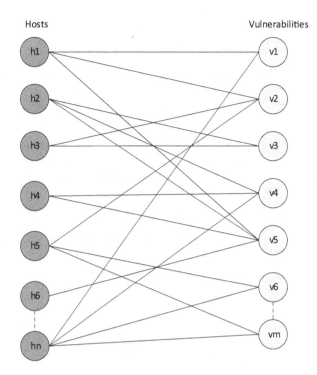

FIGURE 7.13
Graph with non-uniform distribution of n hosts and m vulnerabilities; 3 vulnerabilities per host.

vulnerabilities are represented by a bipartite graph. An anti-coordination game is over the bipartite graph as shown in Figure 7.13 is used to understand the benefit of software diversity. The software diversity is chosen based on Renyi entropy.

The model assumes that there are k vulnerabilities per host $h_i \in H$. A compromised vulnerability thus affects $\frac{nk}{m}$ hosts. Let p_i be the probability that host h, chosen at random, is connected to vulnerability v_i, $p_i = \frac{deg(v_i)}{kn}$. A fraction of edges are connected to each vulnerability. The diversity number N_a is denoted by equation:

$$N_a = (\sum_{i=1}^{m} p_i^m)^{\frac{1}{1-a}}. \tag{7.2}$$

In the equation above, N_a is reciprocal of $(a-1)^{th}$ root of weighted mean of $(a-1)^{th}$ power of p_i. In this equation $N_1 = lim_{a\to1}N_a$., which is the same as Shannon Entropy, i.e., $log(N_1) = H = -\sum_{i=1}^{m} p_i log(p_i)$. As the value of a is increased from $-\infty$ to $+\infty$, the N_a changes weightage assigned from least to most connected vulnerabilities.

TABLE 7.1

Two-Play Anti-Coordination Game Payoff Matrix

	Stay	Switch
Stay	$-c_2, -c_2$	$0, -c_w$
Switch	$-c_w, 0$	$-(c_2 + c_w), -(c_2 + c_w)$

Anti-Coordination Game for two players can be considered in terms of two hosts h_1, h_2 with vulnerabilities v_1, v_2. Each host can stay with vulnerability v_1 or switch to vulnerability v_2. The switching cost c_w is the cost of user getting familiar with a new version of software.

If both hosts h_1 and h_2 plan to stay with original software, the cost incurred due to attack would be c_2, and $c_w < c_2$. The intrinsic cost of one vulnerability is c_0. There payoff matrix for two-player anti-coordination game can be seen in Table 7.1.

The game consists of two pure strategy Nash Equilibrium, i.e., (*Stay, Switch*) and (*Switch, Stay*) and a mixed strategy Nash Equilibrium with cost $\dfrac{c_2 + c_w}{2c_2}$.

This indicates that in a large network some hosts may choose to stay with their original software. An effective diversity technique based on anti-coordination game analysis and Renyi entropy would be to reduce the overlap of vulnerabilities across different hosts.

7.4.4 Markov Game-based MTD

Markov game-based MTD modeling [178] is used for analyzing security capacity of the system and the attack strategy employed by the adversary for defeating the MTD strategy.

An MTD in Markov model is defined as set of defender moves $D = \{\epsilon, d_1, d_2, ...\}$ and attacker moves $A = \{\epsilon, a_1, a_2, ...\}$. The symbol ϵ denotes *No Move*. An algorithm G outputs triple $\{(D_i, A_i, w_i)\}_{i \geq 1}$, where $w_i \in \{0, 1\}$ is used to indicate whether attacker has beaten the defender $w_i = 1$ at time step i or defender has beaten the attacker $w_i = 0$ at time step i. The MTD algorithm 'M' makes use of coin toss to check the attacker's and defender's adversarial moves in next state along with the order of the moves.

If we consider the game to be ending in state 'k' for the attacker or defender, if either of them enters state 'k' during first T time slots. A $(k + 1) \times (k + 1)$ matrix M (Markov chain state transition matrix) can be used to show attackers and defenders moves. If the attacker is currently in state 'i', and decides to take an action which transitions him into the state 'j', the transition probability of the move is denoted by M_{ij}.

An M-MTD (M^D, λ, M^A, μ) can be defined by:

- Parameters λ and μ, $0 \leq \lambda + \mu \leq 1$, which represent the rate of play for attacker and defender.

- $(k+1) \times (k+1)$ transition matrices M^A and M^D; for $i, j \in \{0, ..., k\}$. M_{ij}^A and M_{ij}^D denote the transition probability from state 'i' to state 'j' for attacker and defender, respectively.

A three-valued coin flip decides who moves in a particular state. The attacker moves with probability λ in state 'i', the defender moves with probability μ and with probability $1 - \lambda - \mu$, no one moves in state 'i'.
The MTD game (M^D, λ, M^A, μ) can be fully described using,

$$M = \lambda M^D + \mu M^a + (1 - \lambda - \mu)I_{k+1}, \tag{7.3}$$

where I_{k+1} is the identity matrix in the MTD game.

7.4.4.1 IP Hopping Using Markov Game Modeling

IP hopping refers to continually changing the IP address of the hosts on the network which an adversary is planning to target. The adversary has to continually rescan the network in order to discover the target hosts.

In a TCP communication, the adversary can send SYN packet. If the target host replies back with an ACK, the adversary has successfully discovered the vulnerable host. The defender can select a new IP address from the pool of IP addresses at random, with a rate of λ. The adversary can probe with a rate of μ in order to find the target host. The adversary keeps track of IP addresses already tried in the past states of the game.

The Markov game of $N + 1$ states are shown in Figure 7.14. The state N represents the winning state for attacker or defender. The states $0, 1, ..., N - 1$ represent the states at which IP address was randomized. A state q, $0 \leq q \leq 1$ shows that IP address was randomized exactly q states ago.

- $M_{q,0} = \lambda$. The defender randomizes the IP address with probability λ and the game transitions back to state 0.

- $M_{q,q+1} = \dfrac{\mu(N - q - 1)}{(N - q)}$. The attacker tries IP address with probability μ and with probability $1 - \dfrac{1}{N - q}$, the IP address is incorrect.

- $M_{q,N} = \dfrac{\mu}{N - q}$. The attacker tries IP address with probability $\dfrac{1}{N - q}$. The IP address is correct for attacker and the attacker wins the game.

- $M_{q,q}$. Both attacker and defender do nothing with probability $1 - \lambda - \mu$.

- $M_{N,0} = \lambda$, $M_{N,N} = 1 - \lambda$. The attacker does not need to try any new IP address. Only if defender randomizes the IP address, the attacker will be kicked out of the winning state.

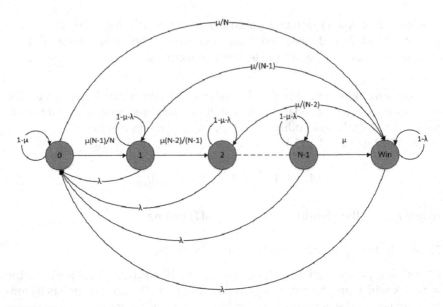

FIGURE 7.14
Single target hiding Markov chain.

For an M-MTD game, stationary distribution $\pi M = \pi$, an initial *state 0*, represents a worst-case start state for an adversary. The probability that the attacker wins after T time steps is $\leq T \times \pi(k)$, where k is the winning state.

7.4.4.2 Winning Strategy for Adversary

The stationary distribution for single target hiding for $1 \leq q \leq N-1$, can be computed as

$$\pi(0) = (1 - \mu)\pi(0) + \lambda\pi(N) + \lambda \sum_{q=0}^{N-1} \pi(q), \tag{7.4}$$

$$\pi(q) = (1 - \lambda - \mu)\pi(q) + \mu \frac{(N - q)\pi(q - 1)}{N - q + 1}, \tag{7.5}$$

$$\lambda(N) = (1 - \lambda)\pi(N) + \sum_{q=0}^{N-1} \mu \frac{\pi(q)}{N - q}. \tag{7.6}$$

The recurrence relation in Equation 7.5 can be solved by,

$$\pi(q) = \frac{\mu}{\mu + \lambda} \frac{N - q}{N - q + 1} \pi(q - 1), \tag{7.7}$$

$$\pi(q) = (\frac{\mu}{\mu + \lambda})^q . \frac{N - q}{N} \pi(0). \tag{7.8}$$

From Equation 7.4, $\sum_{q=1}^{N} \pi(q) = 1 - \pi(0)$, can be solved by $\pi(0) = \frac{\mu}{\mu + \lambda}$,

$$\sum_{q=0}^{N-1} \frac{\pi(q)}{(N - q)} = \frac{1 - \left(\frac{\mu}{\lambda + \mu}\right)^N}{N}. \tag{7.9}$$

Substituting in Equation 7.6, we get the winning state for the adversary:

$$\left\{1 - \left(\frac{\mu}{\lambda + \mu}\right)^N\right\} \frac{\mu}{\lambda N}. \tag{7.10}$$

The amount of time/cost spent by an adversary in defeating a single target hiding strategy can thus be modeled as MTD Markov Game. The defender can use these values to provision optimal MTD in order to increase attack cost for the adversary.

7.5 Evaluation of MTD

The study of MTD effects, when applied to the network can serve as a guide for the deployment of various MTD solutions. The costs and benefits associated with quantifiable security and usability metrics can help in the measurement of effectiveness. In this section, we consider the frameworks that measured the effectiveness of MTD.

7.5.1 Quantitative Metrics for MTD Evaluation

The metrics calculated from cybersecurity testbed serve as a good measure for MTD operational deployment success. A quantitative framework based on Cyber Quantification Metrics (CQM) has been discussed by Zaffarano et al. [294]. The testbed utilizes four key metrics, i.e., *Productivity, Success, Confidentiality* and *Integrity* for both attacker and defender.

In Table 7.2, $\mu = < T, A >$ is the average of duration attributes over the tasks defined by the set A. The valuation of a particular task is denoted by v. The cost of deploying MTD is the valuation of mission productivity valuation with and without MTD. Table 7.2 depicts various metrics and their meaning from the attacker and defender perspective.

TABLE 7.2

Cyber Quantification Metrics for MTD Evaluation

Metric	Defender	Attacker	Formula
Productivity	Rate of task completion	Rate of exploitation	$Prod(\mu, v) = \frac{1}{|T|}$ $\sum_{t \in T} v(t, duration)$.
Success	Normalized value of task completion success	Normalized value of attack success	$Success(\mu, v) = \frac{1}{|T|}$ $\sum_{t \in T} v(t, succ) \in [0, 1]$.
Confidentiality	How much information is exposed by a task/service	Attacker activity fraction that can be detected	$Con(\mu, v) = \frac{1}{|T|}$ $\sum_{t \in T} v(t, unexp)$.
Integrity	Fraction of information produced by task which is preserved	The accuracy of information viewed by the attacker	$Int(\mu, v) = \frac{1}{|T|}$ $\sum_{t \in T} v(t, intact)$.

7.5.2 MTD Analysis and Evaluation Framework

MTD Analysis and Evaluation Framework (MASON) considers the problem of multi-stage attacks in the SDN environment. The security budget for deployment of MTD countermeasures in a network can be limited, thus analysis of most critical services in the network from centrality and security perspective is very important. In this research work, the authors use system vulnerability information and intrusion detection system (IDS) alerts to compose a threat score for network services.

Snort IDS and the Nessus vulnerability scanner are used for intrusion detection and vulnerability scanning in the network. The logs are collected from both these agents into a centralized Elastic Search, Log Stash, Kibana (ELK) framework.

A page rank-based threat scoring mechanism has been used in this work for combining static vulnerability information and dynamic attack events. The SDN controller Opendaylight (ODL) as shown in the Figure 7.15 interacts with ELK framework, in order to select nodes with high threat score for MTD countermeasure port hopping. The experimental results suggest that deployment of MTD countermeasure on top 40–50% of the services can reduce the cumulative threat score of the network by almost 97%.

The framework does not consider the factors such as capacity to have an impact on network bandwidth, Quality of Service (QoS) while performing MTD-based port hopping. Other MTD evaluation frameworks such as [68] and [85] consider performance, available resources and stability constraints for MTD.

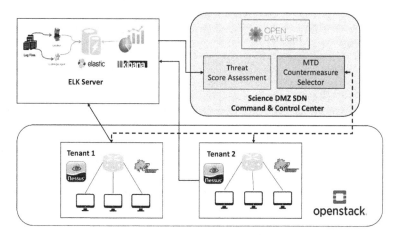

FIGURE 7.15
MASON framework for MTD evaluation.

Summary

Cyber-attacks continue to pose a major threat to the critical infrastructure. There are many reactive defense mechanisms that propose detection, analysis, and mitigation of cyber-attacks. The static nature of cloud systems and defense mechanisms, however, gives an asymmetric advantage to the attackers. Attackers can perform reconnaissance, identify potential targets and launch the attack at will in such a scenario. MTD has emerged as a potential solution to provide proactive security, by constantly changing the attack surface and reducing the window of opportunity for the attackers. This chapter identified and discussed the state-of-the-art MTD defense mechanisms that shuffle, diversify and randomize the existing infrastructure, software and virtual machines. There are trade-offs associated with each MTD technique such as the impact on service availability, computing resources and security investment required for deployment of scalable MTD solutions, which we have discussed to an extent in this chapter. While MTD does not guarantee absolute security, it certainly reduces the attack surface and makes it difficult for the attacker to achieve desired attack goal. The readers are encouraged to evaluate different MTD techniques discussed in this chapter on various cloud platforms such as Google Cloud, MS Azure, AWS, apart from SDN-managed cloud environments.

8

Attack Representation

One of the key components of network defense is representation and visualization of network attacks. Security provisioning in a large network consists of proper authentication of users, authorization of access to software and hardware resources in the network, proper storage and transfer of data across the network. There are multiple entry points of data ingress and egress in a network. Since the volume of information is huge to be analyzed by a human expert, we need logical and efficient data structures for tracking current and potential future activity of normal users and attackers in the network. This is very critical from the point of Threat Assessment, risk analysis and Attack Containment.

In this chapter, we introduce many state-of-the-art data structures that have been utilized in the representation of security threats and the propagation of attack, which makes security assessment easier for a security administrator. We introduce the cybersecurity metrics that are currently used in industry and research for quantification of security threats in Section 8.1, along with an illustrative example of attack propagation in a network. The qualitative and quantitative metrics discussed in Section 8.1 serve as a motivation of attack graph introduced in Section 8.2, which are some of the most popular data structures for attack representation. We also consider probabilistic analysis and threat ranking using attack graphs in Section 8.2. Another important data structure attack tree has been described in Section 8.3, along with examples of attack propagation represented using an attack tree. Attack trees serve as a foundation for attack countermeasure trees (ACT's) that showcase the countermeasures to detect and mitigate attacks in a network along with a qualitative and quantitative analysis of different detection and mitigation techniques. We discuss some other attack representation models in Section 8.5 and analyze the limitations of different attack representation methods in Section 8.6.

8.1 Introduction

The enterprise networks are becoming large and complex. The vulnerabilities and threats are evolving constantly. The management of network security is a challenging task. Even a small network can have several attack paths that can cause a compromise of key services in the network. The incidence response is many companies is still based on instinct and experience of the security professionals. The means of measuring network security risk are very few and limited. The adage "what can't be measured can't be effectively managed" applies here. The lack of good security measurement and management methods make it a challenging task for security analysts and network operators to measure the security status of their network [58].

8.1.1 Cybersecurity Metrics

The security metrics are tools that can be used to measure the security posture of the security state of the organization. Without security metrics, the security analyst cannot answer security questions, such as:

- How secure is my current network configuration?
- If I change the network configuration, will my network become more secure or less secure?
- How can I plan security investment to have a certain level of security?

The security experts use some informal ways of measuring security. The risk analysis based on network threats, vulnerabilities and potential impact [58] is one such example.

$$Risk = Threats \times Vulnerabilities \times Impact. \tag{8.1}$$

The risk measurement can be formalized by the notion of attack surface. The vulnerabilities measurement tools such as NESSUS use security metrics such as the Common Vulnerability Scoring System (CVSS) [72] to quantify the attack surface. The cyber-threats can be measured and mitigated by reducing the attack surface.

8.1.2 Common Vulnerability Scoring System (CVSS)

CVSS is an opensource framework that defines characteristics of software vulnerabilities. It has been adopted as an industry standard by many security frameworks and attack representation methods.

Figure 8.1 shows various subcomponents comprised by the CVSS metric groups. The CVSS metric measurement is divided into three metric groups: Base, Temporal and Environmental. The Base group is used to measure the intrinsic qualities of a vulnerability. The Temporal group measures the vulnerability characteristics that change over time, and the Environmental group measures the characteristics of a vulnerability that are specific to a user's environment. A Base Score (BS) between {0, 10} is assigned to a

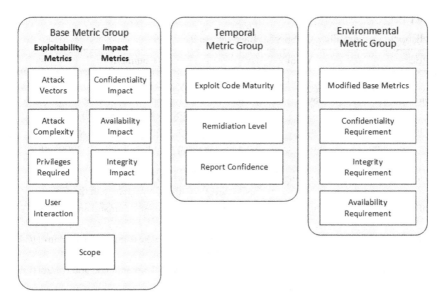

FIGURE 8.1
CVSS v3.0 Metric Groups.

Software Vulnerability and the score can be changed based on Temporal and Environmental metrics. The CVSS score thus captures the principal characteristics of the vulnerability and provides a numerical value that reflects its severity.

8.1.3 CVSS Use Case

CVSS score for a vulnerability is calculated based on several factors as can be seen in Figure 8.1. We analyze a specific example of a vulnerability and corresponding values of CVSS metrics.

The GNU Bourne Again Shell (Bash) vulnerability also known as *Shellshock* has been indexed as CVE-2014-6271 in CVSSv3.0. The older versions of bash ≤ 4.3 allowed trailing strings in the function definition to be part of environment variables. This allowed remote attackers to execute arbitrary code via crafted environment variables. For instance, the attacker can target the Apache HTTPD Server running dynamic content CGI modules. The attacker can craft a request consisting of environment variables. The handler of HTTP requests (GNU bash shell) will interpret requests with the privilege of Apache HTTP process. The attacker can use the privilege to install malicious software, enumerate victim's account details, and perform the denial-of-service (DoS) attack. The values of various metrics which are part of CVSS for Shellshock vulnerability have been showcased in Table 8.1.

TABLE 8.1

Shellshock Vulnerability CVSS v3.0 Base Score: 9.8

Metric	Value	Comment
Attack Vector	Network	Web Server Attack.
Attack Complexity	Low	Attacker needs to access services using bash shell as interpreter or target bash shell directly.
Privileges Required	None	CGI in web server requires no privileges.
User Interaction	None	No interaction required for attacker to launch successful attack.
Scope	Unchanged	GNU bash shell, which is a vulnerable component can be used directly without any change in the scope.
Confidentiality Impact	High	Attacker can take complete command and control (C&C) of the affected system.
Availability Impact	High	Attacker can take complete command and control (C&C) of the affected system.
Integrity Impact	High	Attacker can take complete command and control (C&C) of the affected system.

8.1.4 Attack Scenario Analysis

The data structures used for the representation of information such as current hosts, users, privilege level, vulnerabilities, open ports, etc. should be meaningful and efficient to help network administrators to make useful conclusions regarding current threat level of the network and potential future targets in the network.

Consider a network attack scenario as shown in Figure 8.2. A remote attacker can exploit a security flaw such as weak authentication or unpatched vulnerability on network firewall residing at the gateway. The attacker can in effect gain root-level privilege on firewall. Once the attacker has access to *Internal Network*, in a second stage attack can scan for the machines connected to firewall on *Internal Network*, i.e., *Web Server* running on port *80* and *File Server* running on some other random port.

The attacker can exploit *Buffer Overflow* vulnerability on Web Server and gain root-level privilege on *Web Server*. This will allow the attacker to scan nodes connected to *DMZ Network* and discover *Database Server* on *DMZ Network*. In stage 3 the attacker can exploit *SQL Injection* vulnerability present on *Database Server*. The attacker can ex-filtrate database credentials or corrupt the database.

The series of steps taken by the attacker starting from initial stage and privilege in the network to reach the goal node is known as *Attack Path*. Several data structures such as arrays, 2-D matrices, Binary Decision Diagrams (BDDs), Attack Graphs, Attack Trees, Petri-nets, etc. have been used by security researchers for representation of network attackers path to gain higher privileges in network by exploitation of security flaws and vulnerabilities.

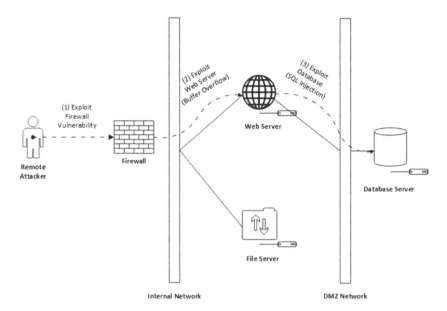

FIGURE 8.2
Attack propagation in a network.

Each data structure has an associated cost in terms of time and space complexity for representation of network attacks.

Attack Graphs (AGs) and Attack Trees (ATs) are two tools that have been extensively used to represent multi-stage, multi-hop attacks in the network. These tools provide a picture of current connectivity, vulnerability information, and potential attack paths in a network. We can leverage information gathered from various sources such as network traffic, system logs, vulnerability scan results, etc., with help of attack graphs and trees to detect and mitigate potential attacks. Attack Representation Methods (ARM) provide time efficient and cost effective network hardening capabilities as discussed by Jajodia et al. [134].

8.1.5 Qualitative and Quantitative Metrics

Attack Graphs present a qualitative view of the security problems, e.g., what are possible attacks in the system and what attack paths can be taken by the attacker to reach the target node in the network.

CVSS metrics, on the other hand, can be used to interpret quantitative information about the vulnerabilities such as complexity of a network attack, impact on confidentiality or integrity of the system if the exploit is successful, etc.

Component Metric can be generated using the numeric value of the CVSS metric. The Conditional Probability of success of an attack can be represented using Component Metric. For instance an attacker needs c_1(network access) to launch an attack on the vulnerability, and c_2(host compromised) is the consequence of the successful exploit, i.e., $Pr[c_2 = T | c_1 = T]$.

Cumulative Metric is used to represent the aggregate of probabilities over the attack graph. Suppose a dedicated attacker tries exploiting all possible attack paths in the system, cumulative metric denotes the probability of attack success along at least one attack path.

8.2 Attack Graph

An Attack Graph $G = \{N, E\}$ consists of nodes (N) and edges (E). The node set can be represented as:

- The nodes of attack graph can be denoted by $N = \{N_f \cup N_c \cup N_d \cup N_r\}$. Here N_f denotes primitive/fact nodes, e.g., *vulExists(Web-Server, BufferOverflow)*, N_c denotes the exploit, e.g., *execCode(Web-Server, apache)*, N_d denotes the privilege level, e.g., *(user, Firewall)* and N_r represents the root or goal node, e.g., *(root, DatabaseServer)*;

- The edges of the attack graph can be denoted by $E = \{E_{pre} \cup E_{post}\}$. Here $E_{pre} \subseteq (N_f \cup N_c) \times (N_d \cup N_r)$ ensures that preconditions N_c and N_f must be met to achieve N_d and $E_{post} \subseteq (N_d \cup N_r) \times (N_f \cup N_c)$ means post-condition N_d achieved on satisfaction of N_f and N_c.

Definition 1 (*Initial Conditions*) *Given an attack graph G, initial conditions $N_i \subset (N_f \cup N_c)$ refer to nodes in the attack graph used by an attacker as a starting point with the goal of exploiting root node N_r. All other conditions that are a result of some exploit and cannot be disabled without removing exploit conditions are known as intermediate conditions $N \setminus N_i$.*

The attack graph has two different kinds of representation structures, the ovals represent the primitive nodes and exploit nodes, i.e., $N_f \cup N_c$. The diamonds N_d represent the exploit nodes that are a result of satisfaction of preconditions N_f and N_c.

The network in Figure 8.3 consists of web server directly accessible via the Internet. The backend database server consists of sensitive data like employee identification numbers, company records, etc. Both the web server and user subnet can access the database server. The web server is affected by vulnerability CVE-2010-3947. The remote attacker can discover this vulnerability by scanning the network and execute malicious code on the web server.

The database server consists of SQL Injection vulnerability CVE-2009-2106 which allows privilege escalation for a normal user. The workstations have an outdated version of Internet Explorer (IE). This software is affected by

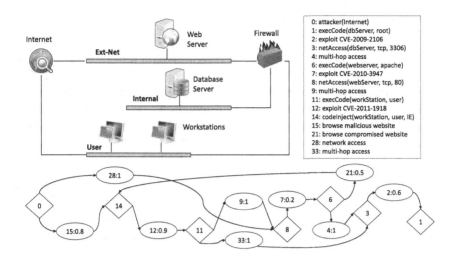

FIGURE 8.3
Attack graph.

vulnerability CVE-2011-1918. The attacker can create an infected website. The workstation user accessing this website using IE can be compromised. The right-hand side of Figure 8.3 shows labels associated with network vulnerability and reachability information.

The attack graph as shown in the Figure 8.3 above can be constructed using tools such as MulVAL [211]. The initial privilege of attacker located on the Internet is represented by *Node 0*. When pre-conditions of an exploit are met, an attack can be accomplished. The pre-conditions are joined using logical AND relation. The exploit of web server *Node 7 (7:0.2)* in attack graph can occur if an attacker can access the web server via TCP port 80 *Node 8*. If a postcondition node in the attack graph has multiple incoming arcs, this means that a privilege can be obtained in more than one way. The incoming arcs into diamond nodes are joined by logical OR relation.

For instance network access to database server *Node 3* can be obtained by web server *Node 6* or workstation *Node 11*. The attack graph can help network admin to identify multiple attack paths that can lead to a compromise of various hosts. A multi-hop attack can be launched by the remote attacker to first exploit web server vulnerability and use it as a stepping stone to access database server *(0, 28, 8, 7, 6, 4, 3, 2, 1)*. Another attack scenario involves attacker tricking workstation user to click on a malicious website and use the workstation as a stepping stone to access database server *(0, 15, 14, 12, 11,…)*.

There can be an exponential number of attack paths in worst-case scenarios in an attack graph. The attack graph in this example has been constructed using known vulnerabilities. However, we can also simulate artificial vulnerabilities as input for the attack graph. This can be used for reasoning about *Advanced Persistent Threat* scenarios.

8.2.1 Probabilistic Attack Graphs

Security metrics in attack graphs help in assessing the risk likelihood of network configuration. Many attack graph representation techniques focus on individual security vulnerabilities. The problem with this approach is that multi-stage network attacks exploit the dependencies between security vulnerabilities. Another issue in attack representation techniques is the fact that they consider binary representation of network security *secure* or *insecure*. This is limited because it is desirable for a security administrator to find a relatively secure option amongst security configurations.

Probabilistic attack graphs utilize security metrics based on existing vulnerability scoring systems such as *Common Vulnerability Scoring System (CVSS)* [72] and attack graph model. The model assigns probabilistic values to the network vulnerabilities, i.e., likelihood of a vulnerability to be exploited. Additionally, the model incorporates casual dependencies between the network vulnerabilities.

The security metrics for attack graph can use *priori probability* value for nodes. The probability value indicates the likelihood of a threat becoming active and difficulty of vulnerability exploitation. We can consider prior Risk Probability, denoted by G_v for the root node $N_r \subseteq N_d$. Probability value of each internal risk node $e \in N_c$ is denoted by $G_m[e]$. This value is calculated using Base Score (BS) from CVSS metric. Base score is a combination of exploitability and Impact-Vector (IV) as shown below:

$$BS = (0.6 \times IV + 0.4 \times E - 1.5) \times f(IV)$$
$$IV = 10.41 \times (1 - (1 - C) \times (1 - I) \times (1 - A)), \qquad (8.2)$$
$$E = 20 \times AC \times AU \times AV;$$

$$f(IV) = \begin{cases} 0 & if \quad IV = 0. \\ 1.176 & otherwise. \end{cases} \qquad (8.3)$$

The impact value (IV) is calculated using security parameters like *confidentiality (C)*, *integrity (I)* and *availability (A)*. The exploitability score E is composed of *access vector (AV)*, *access complexity (AC)* and *authentication instances (AU)*. The value of BS from CVSS metrics lies between 0 and 10. We normalize the value between (0,1] for probabilistic attack graphs:

$$Pr(e) = \frac{BS(e)}{10} \quad \forall e \in N_c. \qquad (8.4)$$

The likelihood of attack propagation depends on the conjunct and disjunct relation between exploits. The Risk Probability of current node is determined using *Conditional Probability*. This value depends upon Risk Probability of predecessors and their relationship with the current node.

- If we consider the predecessor set $W = parent(n)$ of any attack step node in attack graph $n \in N_c$, conditional Risk Probability of attack step node is given by

$$Pr(n|W) = G_m[n] \times \Pi_{s \in W} Pr(s|W).$$ (8.5)

- On the other hand, if the predecessor node is having disjunctive relationship with predecessor set $W = parent(n)$, the Conditional Probability value is given by

$$Pr(n|W) = 1 - \Pi_{s \in W}(1 - Pr(s|W)).$$ (8.6)

Once the Conditional Probability values of internal/step nodes have been calculated, we can merge them to calculate the cumulative or absolute probability values according to the equations below.

- For the attack step nodes $n \in N_c$ with predecessor set $W = parent(n)$,

$$Pr(n) = Pr(n|W) \times \Pi_{s \in W} Pr(s).$$ (8.7)

- For the attack privilege nodes $n \in N_c$ with predecessor set $W = parent(n)$,

$$Pr(n) = 1 - \Pi_{s \in W}(1 - Pr(s)).$$ (8.8)

The values of cumulative probability can be used by the network administrator for devising effective security countermeasure strategies.

8.2.2 Risk Mitigation Using Probability Metrics

There are several factors that need to be considered when designing a defensive strategy based on network attack graphs such as money, time, and impact on the service availability. We can use Risk Probability calculation based on the approach proposed by Homer et al. [114].

Table 8.2 shows the impact of various countermeasures on the attack graph depicted in Figure 8.3, i.e., patching vulnerability on DB Server, Web Server and workstation firewall (Change in network access). *Column 2* shows the initial values of probability for the attack to be successful.

The analysis of various countermeasures in terms of cost-intrusiveness can be performed by comparing values of attack success with each countermeasure in *columns 3-6*. Patching Web Server can eliminate the risk of compromise on Web Server, but there is limited impact on the values of Database Server and Workstations. Additionally, Web Server may not contain sensitive information so protecting Web Server may not be the best option. The attacker can target workstation and use it to access database server in a multi-hop attack.

TABLE 8.2

Risk Mitigation for Attack Graph

Host	Attack Prob.	Patch Web Server	Patch DB Server	Patch Workstation	Network Access Change
DB Server	0.47	0.43	0	0.12	0.12
Web Server	0.2	0	0.2	0.2	0.2
Workstations	0.74	0.74	0.74	0	0.74

The Database Server can be patched to eliminate the risk of losing important information, but the downtime associated with patching the Database Server may impact the business. Also, an attacker may still target workstations, as patching Database Server does not help in securing the workstations. The patch on the workstation can eliminate the attack path to the Database server, thus eliminating the risk to attack Database Server. The attacker can still target the Web Server, but the Database Server is secured.

The Firewall option, last column in Table 8.2, can be considered as a countermeasure, to block access from Workstations to Database and Web Server. This will help reduce the security risk of possible compromise of important assets in the network. This can be considered a viable option depending on the network setup, impact on business and other security and usability constraints. Depending upon the cost-benefit analysis, a network administrator can consider one or more countermeasures to deal with attack scenarios described using an attack graph.

8.2.3 Attack Graph Ranking

The states in the attack graph can be assigned a ranking, which indicates the probability of an intruder reaching the state. Mehta et al. [189] propose ranking algorithm based on these state probability values. The scheme utilizes ranking scheme similar to Google's *Page Rank* algorithm [212]. The PageRank algorithm, which is based on user behavior, assumes there is a "random surfer" who visits web pages until he gets bored. Once the random surfer gets bored, he can click on any random page. To capture the notion of randomness, a damping factor d is used, where $0 < d < 1$. The factor $1 - d$ denotes random transitions from a given state to all possible states in a web graph consisting of web pages as graph nodes and web links as edges.

If the graph consists of N nodes (web pages), Let In(j) be set of pages linking to web page j and Out(j) be set of out-links from page j. The probability of random surfer being at page i is given by

$$\pi_i = \frac{1-d}{N} + d \sum_{j \in In(i)} \frac{\pi_j}{|Out(j)|}. \tag{8.9}$$

Let PageRank vector be $R = (r_1, r_2, r_3, \ldots, r_N)^T$, where rank of page i is r_i. The PageRank of page i is defined by the probability π_i. Equation 8.9 is computed recursively until π_i converges. In the attack graph ranking, the event probabilities are assigned to attack graph states, and the probability for the system is computed until the values converge to a steady state. The parallels between PageRank random surfer model and using a similar approach in attack graph ranking is justified by the fact that brute force methods such as Distributed Denial of Service (DDoS) use sequential probing of various network services. Automated attacks like viruses, on the other hand, behave randomly.

8.3 Attack Tree

Attack Tree [240] is another method of representing system security. The Attack Tree represents the network attacks. Attack Tree represents a monotonic path taken by an attacker starting from a leaf node to reach a goal node. Attack Tree usually consists of set of *AND* nodes and *OR* nodes. The *OR* nodes represent one or more ways in which a goal node can be reached, whereas *AND* nodes represent different conditions that must be fulfilled in order to achieve a goal node. Children of the node are refinements of this goal, and the attacks that can no longer be refined are represented by leaf nodes [181].

Consider the Attack Tree [240] in Figure 8.4. The triangle represents *OR* node, the logic *AND* gate symbol represents the *AND* node, and the rectangle box represents the *leaf* node. The *Password Hash* can be obtained by attacker from *Shadow File OR Readable Password File*.

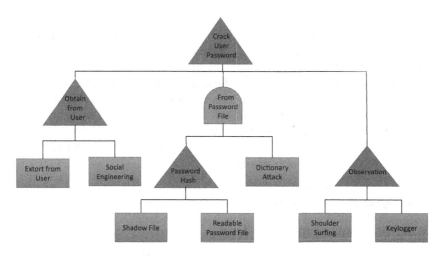

FIGURE 8.4
Attack tree example.

The attacker can obtain *Password File* using *Password Hash AND* textit-Dictionary Attack. Similarly, password can be obtained from user by *Extortion OR Social Engineering*. Both these techniques in addition to other methods like *Observation* can be used to crack a user password. An attack tree represents all relations specified by OR, *AND* conditions. The different combinations of tree nodes (attack suites) can help an attacker reach his desired goal. We formally define attack components and suites below.

Definition 2 *Attack component set can be defined as C. An attack is finite, non-empty multi-set of C and attack suite is finite set of attacks. We can denote universe of attacks by $A = M^+(C)$ and universe of attack suites as $S = P(A)$.*

For example, if we have $C = \{$weak credentials, buffer overflow, root access$\}$, then attack suites for gaining root access on a system can be defined as, $\{\{$weak credentials, root access$\}$, $\{$buffer overflow, root access$\}\}$.

The nodes can be assigned different values in order to calculate the attack or defense cost, e.g., easy to exploit and difficult to exploit, intrusive and non-intrusive. The cost of attack/defense can be quantified in terms of time required for exploitation, the cost in terms of dollar amount, etc. For instance, the time required to obtain password directly from the user may be lesser than using dictionary attack and password hash.

Definition 3 *An Attack Tree [181] can be defined by three tuple $(N, \rightarrow, N_r)$*

- *N is all possible nodes in the tree;*
- *$S^+(N)$ is multi-set of all possible subsets of nodes N;*
- *$\rightarrow \subseteq N \times S^+(N)$ denotes transition relation;*
- *N_R represents the goal node of the attack tree.*

For the example in Figure 8.4, $S^+(N) = \{\{$*Obtain from User, Crack Password*$\}$, $\{$*From Password File, Crack Password*$\}$, $\{$*Observation, Crack Password*$\}\}$. The goal node is $N_r = \{$*Crack Password*$\}$.

If we have more than one value of the same node, it is possible that calculation of attack cost is not unique and consequently undefined. In Figure 8.5, the attacker needs to invest *5 minutes* and *10 dollars* in case of leaf node on the left, whereas in case of leaf node on the right investment required for an attack is *10 minutes* and *5 dollars*. The attack profile of cheapest attack in the shortest time is undecidable in this example.

8.4 Attack Countermeasure Tree

Existing state-space models used to represent Attack Trees and counter-measures suffer from scalability challenges. Attack Response Tree [234] is

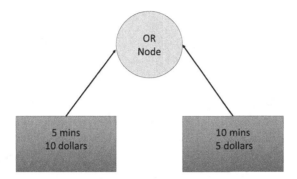

FIGURE 8.5
An undecidable case in attack tree.

used to represent the countermeasure/response (R_1, R_2) for each attack state (A_1, A_2).

As can be seen in Figure 8.6 such a state space model can lead to state space explosion. The number of states is equal to $2^{Leaf\ Nodes}$. Attack Countermeasure Tree (ACT) [235] takes a non-state-space representation to the security modeling. The defense mechanism can be placed at any node in the Attack Tree as opposed to only at the leaf node in the Attack Response Tree. The ACT can be used to perform a probabilistic analysis of system risk, Return of Investment (ROI), attack impact, etc.

Definition 4 *(ACT) can be defined as $ACT = \{V, E, \phi\}$. A_i indicates an attack event, D_j indicates a detection event, M_1 indicates mitigation event and CM_k denotes a countermeasure event.*

- $V = \{\forall k v_k \in A_i \| D_j \| M_l\}$,
- $\phi = \{\forall k \phi_k = \{AND,\ OR,\ k\text{-}of\text{-}n\ gate\}\}$,
- $E = \forall k e_k \in \{(v_i,\ \phi_j) \| (\phi_i,\ \phi_j)\}$,
- $X = (x_{A_1} x_{A_2}..x_{D_1} x_{D_2}...x_{M_1} x_{M_2}..)$ *is state vector of ACT.*

Figure 8.7 shows a different combination of attack, detection and mitigation events in ACT. Figure 8.7(a) shows ACT needs detection event outcome to be *False* along with network attack to achieve goal *Attack Success*. Similarly both detection and mitigation events should be *False* in order for attack to be successful in Figure 8.7(b).

8.4.1 ACT Qualitative and Quantitative Analysis

Qualitative Analysis considers the Mincuts that represents the attack scenarios leading to successful attack. In the given example in Figure 8.8, the goal of

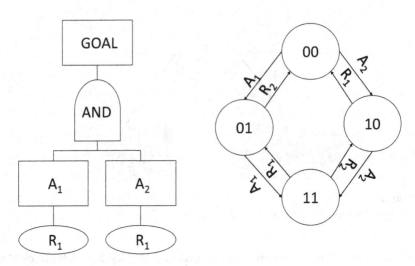

FIGURE 8.6
Attack response tree.

attacker is resetting the BGP session. The mincuts of attack tree (AT) are $\{(A_{111}, A_{12}), (A_{1121}, A_{12}), (A_{1122}, A_{12}), (A_{1123}, A_{12}), (A_2)\}$.

Figure 8.9 shows Attack Countermeasure Tree (ACT) for BGP reset attack. The goal node for the ACT can be analyzed from a qualitative and quantitative aspect.

The attack methods for each attack, e.g., A_1 *send reset message to the router* has corresponding detection and mitigation methods. The detection method

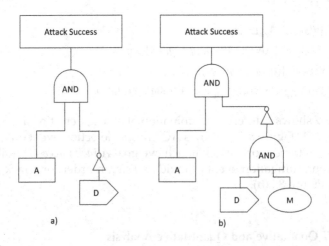

FIGURE 8.7
a) ACT with one attack and one detection event; b) ACT with one attack, one detection and one mitigation event.

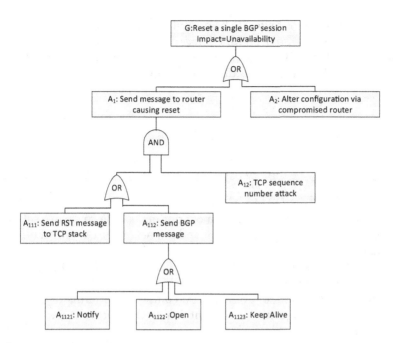

FIGURE 8.8
Attack tree (AT) for BGP reset event.

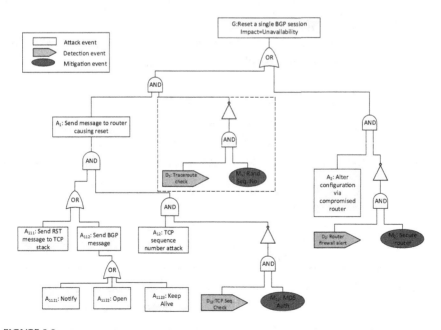

FIGURE 8.9
Attack countermeasure tree for BGP reset event.

for A_1, i.e., D_1 is *Trace route check* and mitigation methods M_1 is *Randomize Sequence Number*. The boolean function $\overline{\phi(X)}$ is used to represent the complement of leaf nodes in ACT. The state of ACT can be depicted using state vector X, where $x_{A_i} = 1$ if an event occurs and 0 otherwise.

Using the mincut set from AT in Figure 8.8, we have,

$$\overline{\phi(X)} = x_{A_{111}}x_{A_{12}} + x_{A_{1121}}x_{A_{12}} + x_{A_{1122}}x_{A_{12}} + x_{A_{1123}}x_{A_{12}} + x_{A_2}. \qquad (8.10)$$

The mincuts for ACT are $\{(A_{111}, \overline{CM_1}, A_{12}, \overline{CM_{12}}), (A_{1121}, \overline{CM_1}, A_{12}, \overline{CM_{12}}),$ $(A_{1122}, \overline{CM_1}, A_{12}, \overline{CM_{12}}),$ $(A_{1123}, \overline{CM_1}, A_{12}, \overline{CM_{12}}),$ $(A_2, \overline{CM_2})\}$, (where $\overline{CM_1} = \overline{D_1 M_1}, \overline{CM_{12}} = \overline{D_{12}M_{12}}, \overline{CM_2} = \overline{D_2 M_2}$).

The mincuts can be used to find most optimal countermeasure set for the ACT, based on the cost of deployment and impact on the overall security state of the network due to a particular detection and mitigation methods that are part of countermeasure (CM). For instance CM_1 and CM_{12} can be used to prevent the attack events A_{1121} and A_{12}.

Quantitative Analysis of ACT can be computed in terms of the Cost of Attack (COA) and security investment required to prevent the attack, i.e., Return of Investment (ROI).

The cost of attack depends upon the input gate for the attack in the ACT.

In the case of AND gate, cost of attack is a sum of the cost of all input events as shown in Table 8.3, whereas in case of OR gate, the attack cost will be a minimum value of the cost of all attack events. The COA for a k-of-n gate is the sum of k lowest attack event cost.

COA or $C_{attacker}$ is a minimum value of attack cost of mincuts. The COA for an attacker in case of BGP reset attack depicted in Figure 8.8 is

$$C_{Attacker} = min\{(c_{A_{111}} + c_{A_{12}}), (c_{A_{1121}} + c_{A_{12}}), (c_{A_{1122}} + c_{A_{12}}), (c_{A_{1123}} + c_{A_{12}}), c_{A_2}\}. \qquad (8.11)$$

The ROI can be computed using values I_{goal} and P_{goal} in addition to cost of attack computed from Table 8.3. I_{goal} depicts the damage to the system if attack is successful. P_{goal} is used to quantify the probability of attack being successful.

TABLE 8.3

Cost of Attack in ACT

Gate	Cost of Attack
AND gate	$\sum_{i=1}^{n} c_{A_i}$
OR gate	$min._{i=1}^{n} c_{A_i}$
k-of-n gate	$\sum_{i=1}^{n} c_{A_i}$

We define system risk as $Risk_{sys} = I_{goal} \times P_{goal}$.

$$ROI = \frac{Risk_{sys}}{C_{attacker}}. \tag{8.12}$$

The defense strategy for a given network can be formulated by considering qualitative and quantitative factors such as mincuts, ROA and ROI associated with ACTs.

8.5 Other Attack Representation Models

8.5.1 Fault Tree

Fault tree [234] is an analytic model that is used to check the state of the system. The root node of the tree is specified and the system is analyzed for undesired operations and events to find subtrees with nodes/leaves that contribute to the event. The fault events are linked by logic gates *AND, OR,* etc.

The Fault tree in Figure 8.10 shows causes at leaf node such as the existence of network connectivity and Software Vulnerability that leads to fault *Privilege Escalation*. Each of the faults connected by *OR* gate can cause *top level event*, in this case *Email Server* crash.

8.5.2 Event Tree

An event tree [234] is used to identify event such as a security breach in the network. The logic used in the event tree is different from the fault tree in

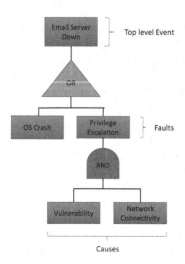

FIGURE 8.10
Fault tree example.

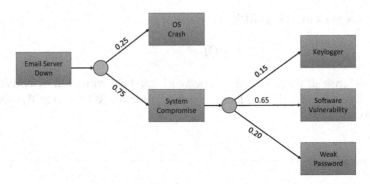

FIGURE 8.11
Event tree.

the sense that event trees do not include decision points requiring logical operators *AND* and *OR*. Event trees use binary logic, in which an event has or has not happened.

The event tree [234] in Figure 8.11 shows the probabilistic values of occurrence of various events leading to *Email Server* crash. The Cumulative and Conditional Probability values of occurrence of each event can be calculated using probability values at each level of the event tree. For instance, the probability that *Email Server* is down due to *Software Vulnerability* is $p(SoftwareVulnerability) \times p(SystemCompromise)$, i.e., 0.4875.

8.5.3 Hierarchical Attack Representation Model

The graph attack representation model (ARM) has issues such as state space explosion caused by an exponential number of attack states. On the other hand, tree-based representation models fail to capture attack path information accurately. Moreover, the changes in network topology and service vulnerability information over time make it difficult to have just a single ARM. Hierarchical attack representation model (HARM) [117] segregates the network topology and vulnerability information by using a different data structure for each. The network topology information is represented using an attack graph (upper layer), whereas the vulnerability information is represented using an attack tree (lower layer).

An update in the network such as host addition, host removal, and vulnerability patching cause a modification in representation structure. The computational complexity for modification afforded by HARM [117] is better compared to structures such as attack graph and attack tree. Additionally, use of a layered approach helps to improve the scalability of HARM. Attack model construction can be parallelized at each layer using HARM. Security and performance analysis on each layer in HARM can be performed separately.

The HARM corresponding to attack scenario in Figure 8.12 shows that an upper layer, attack has network connectivity with *Web Server* and *Database*

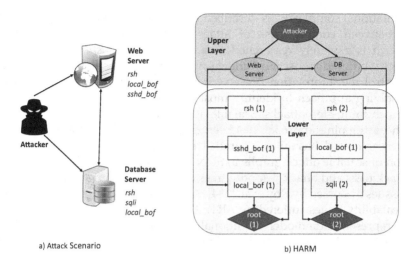

FIGURE 8.12
Hierarchical attack representation model (HARM).

Server represented by attack graph in Figure 8.12(b). At the lower layer, the vulnerabilities on Web Server and Database Server - local buffer overflow (*local_bof*), remote shell (*rsh*) - are represented using an attack tree.

8.6 Limitations of Attack Representation Methods

The attack representation methods have suffered from scalability issues. The model checking approach proposed by Ammann et al. uses counterexamples as a means to check the security policies [28]. This approach will suffer from scalability issues, as path explosion is often an issue with model checking. Ammann et al. performed attack graph-based analysis of network attack chains [29]. The computation complexity for the algorithm is $O(n^3)$ for the worst case attack paths.

MulVAL has utilized security analysis based on logic programming. The research work uses a first-order logic-based modeling approach to identify logical dependencies between attack goals and the configuration information [211]. The attack graph scales polynomially in terms of network size. The running time of MulVAL is estimated to be between $O(n^2)$ and $O(n^3)$. Homer et al. [115] utilize data reduction and attack grouping methods for improving the visualization of attack graphs. The portions of the attack graph not critical for security assessment are trimmed, whereas the similar attacks are grouped in attack graph to improve the understanding of attack goal.

Ingols et al. [131] proposed multi-prerequisite graph-based security assessment. The attack graph scales linearly with the network size. The computational complexity as described in the paper is $O(E + NlgN)$. The balance between completeness of attack representation and scalability of approach is a widely researched area. *Attack Graph distillation* [126] uses severity metrics to choose most critical attack paths. This helps the administrator control the information presented. The paper utilizes a Minimum-Cost SAT solving approach to identify most critical attack paths for the attacker to launch multi-step attacks. Scalable attack graph generation using hypergraph partitioning approach has been discussed by Chowdhary et al. [64]. The research work has utilized parallelized attack graph partitioning for generation of scalable attack graphs in SDN-based cloud network.

Attack trees and attack response trees (ART) suffer from state space explosion issues discussed by Roy et al. [234]. ACT's have been proposed to address the scalability issues suffered by ART. Authors [236] have used a non state-space representation model. Additionally, optimization under different budget constraints has been considered in this work.

Summary

Cybersecurity Metrics reflect the quantitative information about the security state of the network. CVSS is the most commonly used security metric measure. The CVSS score assigned is based on factors such as attack complexity, confidentiality, integrity, availability impact. Attack Graph is a tool to represent the vulnerability, reachability and attack propagation information in a network. The nodes that are required for exploitation of particular vulnerability (cause) are known as pre-conditions. The effect node which is a result of vulnerability exploitation is known as post-conditions. Probability metrics can be used for assessment of attack success probability. Likelihood of attack depends upon the conjunctive and disjunctive relation between the nodes of the attack graph, quantified by conditional and cumulative probability values.

Attack Tree is another representation method that depicts the monotonic path taken by an attacker to reach goal node. The attack tree consists of AND, OR nodes that are required to be fulfilled for state transition. Attack Countermeasure Tree (ACT) represents a non-state-space view of security in a network. ACT also consists of possible defensive and mitigation methods that can help is thwarting an attack. A mincut-based analysis of various attacks which are part of attack tree, and countermeasures (defense and mitigation methods) can be used for qualitative assessment of ACTs. A quantitative assessment requires estimation of attack probability, the impact of exploit and cost of attack for the attacker. Attack representation methods - Attack Graphs and Attack Trees-often suffer from the problem of state space explosion; representation methods with low computation complexity should be selected to handle scalability challenges.

9

Service Function Chaining

There has been a shift in paradigm from the host-centric model to the data-centric model. The network services and computation capacity are available closer to the users. The cloud and data center network and cloud architecture require flexible network function deployment models.

The advent of technologies such as Network Function Visualization (NFV) allows Cloud Service Providers (CSPs) to provision Virtual Network Functions (VNFs) as opposed to physical networking infrastructure, e.g., a traditional router can be replaced by a virtual router. In a traditional network, a sequence of steps is required when connecting various hardware components responsible for handling the traffic coming into and going out of the network.

Service Function Chaining (SFC) is one way of providing end-to-end network connectivity while maintaining different Service Level Agreements (SLAs) promised by CSPs. In this chapter, we discuss different SFC objectives, design and deployment considerations and challenges faced by CSPs in SFC provisioning. Keeping in line with the main theme of the book, one section has been dedicated to security provisioning using SFC framework. The motivation for SFC along with high-level design goals has been discussed in Section 9.1. The SFC architecture, core concepts, and challenges introduced in SFC by topological dependencies, resource availability and configuration complexity have been discussed in Section 9.2. The SFC on top of SDN/NFV framework with different applications such as segment routing has been described in Section 9.3. Section 9.4 introduces different SFC research project-based testbed and their architectures, such as T-Nova and Tacker. The policy composition for SFC deployment has been discussed in Section 9.5. The research works in the field of secured service function chaining such as Secure In Cloud Chaining (SICS) and Network Security Defense Pattern (NSDP) have been discussed in Section 9.6.

9.1 Introduction

The NFV architecture requires an agile service function model, which can handle dynamic and elastic service delivery requirements, handle movement of various service functions and application workload in the network and mechanism to bind the service policies to granular per-subscriber state [223].

TABLE 9.1

Common Causes of Middlebox Failures

	Misconfiguration	Overload	Physical/Electric
Firewall	67.3%	16.3%	16.3%
Proxies	63.2%	15.7%	21.1%
Intrusion Detection System (IDS)	54.5%	11.4%	34%

Middleboxes are the devices used by network operators to perform network functions along the packet's datapath from source to destination, e.g., Web Proxy, Firewall, and Intrusion Detection System (IDS). Researchers have focused efforts on several issues associated with middleboxes such as being easier to use, easier to manage, design and deploy the general-purpose middleboxes for different network functions. A survey of various middlebox deployments conducted by Sherry et al. [249] reveals factors such as increased operating costs caused by misconfiguration and overloads that affect their normal functioning.

As shown in Table 9.1 based on results of a survey [249] of 57 enterprise network administrators, from NANOG network operators group, the misconfigured and overloaded middleboxes are the major reasons of middlebox failure. About 9% administrators reported about six and ten hours per week dealing with middlebox failures. Also, the adoption of new middlebox technology is slow in the industry based on the survey results. In the median case, enterprises update their middleboxes every four years.

Service Function Chaining (SFC) has emerged as a new architecture that helps in better classification of various issues associated with middlebox deployments and their potential solutions. SFC is a capability of network functions (middleboxes) such as firewall, IDS, proxy, and Intrusion Prevention System (IPS) to be connected in form of a chain for traffic steering and end-to-end delivery of packets in a network. Automated provisioning of service chains for network applications with different characteristics can be beneficial from an operational and cost aspect for a CSP. For instance, a VOIP service can be more resource intensive as compared to a proxy service.

A framework for management of SFC can help ensure on-demand provisioning of traffic sessions with user-specified requirements. There has been a surge in traffic of mobile devices and the Internet of Things (IoT), which make deployment of VNFs closer to the user, distributed across multiple data-centric networks even more desirable [46]. SDN is an enabling technology that helps in realizing the VNF capabilities provided by NFV. There are several key assumptions that need to be considered as part of SFC architecture [105]. We have discussed some of them below:

1. The set of service function in a given administrative domain may vary from time to time. There is no global list of service functions that need to be deployed in a particular domain.

2. The policy of SFC and criteria for enabling them is a decision local to the administrative entity of each SF domain.

3. Each administrative entity can define its own SFC logic. There is no global standard on defining the SFC logic.

4. Several SF policies can be applied simultaneously to an administrative domain in order to achieve business objectives.

5. SFC assumes independence of the underlay network setup. The SF architecture places no restrictions on how the connectivity is realized, or the factors such as jitter, bandwidth, and latency affect the connectivity.

9.2 SFC Concepts

In this section, we describe the key terms associated with SFC architecture as discussed in [105].

1. **Network Service** refers to an offering provided by a network service provider. Network service can be composed of single or multiple VNFs, e.g., Firewall, IDS, Load Balancer, etc.

2. **Classification** is used to describe the traffic flow segregation based on local policies defined for a segment of the network. A set of service functions can be assigned to a class of traffic. The element responsible for the classification of the traffic is known as *Classifier*.

3. **Service Function Chain (SFC)** refers to an ordered set of service functions that should be applied to the classified traffic. The order of application of abstract service functions can be sequential or parallel, based on the network requirements, e.g., Firewall and IDS can be used in serial order for SFC. On the other hand, VNF operations such as Monitoring and Firewall can be parallelized as discussed by Sun et al. [262].

4. **Service Function (SF)** is used to describe a function that is used for the treatment of traffic in a certain way. The service function can be implemented using a virtual component, e.g., a virtualized firewall or a physical hardware, e.g., Cisco router or switch. A virtual service function at the network level is also known as Virtual Network Function (VNF). A hardware device can be used to implement one or more service functions.

5. **Service Function Forwarder (SFF)** is a device, physical or virtual, that is used for forwarding traffic to a neighboring section of the network based on the policy defined in the SFC encapsulation. The SFF handles both ingress and egress traffic for a network segment.

6. **Service Function Path (SFP)** is a constrained specification of traversal of packets for a given SFP. The granularity of SFP can be at a level of exact location or it can also be less specific. This allows network operators to have a certain level of control over the selection of SFF/SF.

7. **SFC Encapsulation** consists of information that helps in identification of SFP for SFC aware traffic. In addition, SFC encapsulation may also contain metadata associated with the data plane of the traffic being steered using SFC.

8. **Rendered Service Path (RSP)** refers to the actual path traversed by packets between two endpoints in a network based on the SFP specified by various network operators in each zone as well as SFF/SF visited by traffic along the path.

SFC describes the ordered set of service functions that must be applied in a provided order to offer a service, e.g., QoS service may require SFC containing Proxy and Load Balancer.

The example in Figure 9.1 illustrates the key terms described above. We have two different SFCs based on the application requirement. The SFC parental control blocks certain web traffic not deemed appropriate for

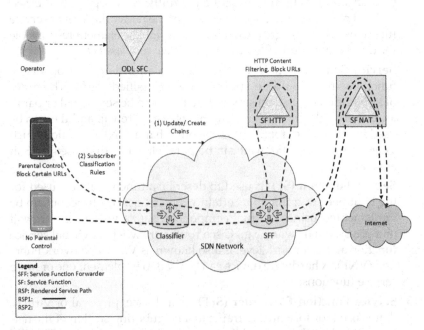

FIGURE 9.1
Service function chaining example.

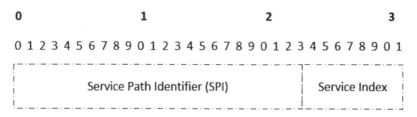

Service Path Identifier (SPI):   24 bits
Service Index (SI):   8 bits

FIGURE 9.2
Network service header.

children. The classifier is used to steer the traffic that belongs to parental control SFC using SFs HTTP and NAT. The inappropriate traffic is filtered using SF HTTP. On the other hand, the SFC with no parental control steers the traffic via SF NAT on the path to the internet. The RSP2 is used to represent the actual traffic path for SFC with parental control in place, whereas RSP1 represents the path traversed by SFC with no parental control. Both service chains traverse through the SFF while accessing the services available on the internet.

Network Service Header (NSH) is a service plane protocol used for the creation of dynamic SFC.

The key components of NSH are the following:

1. *Service Path Identifier (SPI)* as shown in Figure 9.2 is used to identify service function path. This identifier is used by participating nodes in SFC for SFP selection.

2. *Service Index (SI)* in Figure 9.2 is used to identify the location of service function path. The value of SI is decremented as SFs or service function proxy nodes are encountered along the SFP. The default initial value of SI is 255.

3. Optional metadata can be shared between participating entities and SFs.

NSH can help in providing reusable classification of pre-programmed service function paths. Figure 9.3 shows SFC with NSH. The chain information is encapsulated with each packet using NSH. Some advantages and limitations of using NSH as part of SFC have been discussed in Table 9.2.

9.2.1 Challenges in SFC

There are several challenges in deployments of SFC such as Firewall, IDS, and Deep Packet Inspection (DPI) in large-scale environments. Factors like

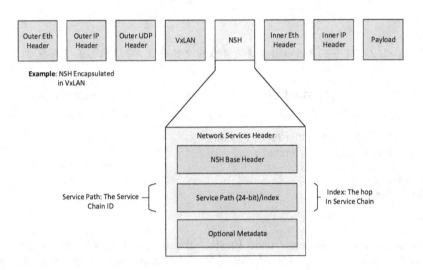

FIGURE 9.3
SFC with network service header.

TABLE 9.2

Network Service Header (NSH) Advantages and Limitations

Advantages	Limitations
NSH allows metadata to be sent as a part of SFC. This can be helpful for optimal service chain deployment	NSH currently has limited support in switches, kernels, and applications
NSH supports a wide variety of underlay and overlay protocols such as MPLS, GRE, VXLAN	SF should be aware of NSH being a part of SFC
NSH helps in simplifying the forwarding complexity due to complex topological dependencies	NSH encapsulation can introduce a delay in the SFC

application delivery, security policies, and network policies are often in conflict with each other when deploying an SFC as discussed by Quinn and Nadeau [224]. We highlight some important issues to be considered for utilizing the full benefits of SFC.

Topological Dependencies: The service delivery offered by SFC is often tightly coupled with network topology. For instance, in cloud services, different tenants communicate via fast tunneling networks such as VLAN. If the service function such as Firewall or IDS is placed in the SFP, the service delivery is affected. This also inhibits the scale, capacity and flexibility across the network.

If the network functions are hardware based or are fixed in place as per network operators design specifications, they are difficult to move around, and the SFC has to impose a strict ordering of SFs for packet delivery.

Configuration Complexity: The logical and physical topology in networks is at times dependent on the order of SFs in SFC. If there is a change in the ordering of SFs, a corresponding physical or logical change is also required in the network. Network operators are hesitant to make such a change due to the fear of the network downtime caused by potential misconfiguration. This leads to static service delivery deployment.

Constrained Availability of SFs: In order to provide high availability based on network topology, redundant service functions are also required in addition to the primary service functions. Topological dependencies can, however, impose constraints on the high availability of SFs.

Ordering and Application of SFs: The administrators have to consider the order dependencies between the SFs. Administrator requires a standard way to enforce and verify the ordering of SFs. There is no standard accepted protocol so far to achieve this objective in SFC. The service policy application in SFC needs topological information, but the available information is not granular enough to be a reliable means for helping in the policy specification.

Transport Dependence: The SFC should be generic enough to support a wide variety of underlay and overlay protocols such as Virtual eXtensible Local Area Network (VxLAN) Ethernet, Generic Routing Encapsulation (GRE) and Multiprotocol Label Switching (MPLS).

Elastic Service Delivery: The resource flexing such as adding and removing network functions based on the real-time traffic needs is hard to achieve in existing networks due to topological dependencies and routing changes that are required to ensure service delivery.

Traffic Selection Criteria: The existing SFC deployments have coarse traffic selection policies. Policy routing and access control filtering can be used to achieve granular traffic selection in SFC.

Security Considerations: (i) The service overlay depends on the transport protocols such as GRE and MPLS in the network. If the security requirement wants confidentiality or authentication in the service delivery, a secured protocol such as IPSec should be selected. (ii) The classification policy used for selection of the underlay or the overlay protocol for SFC must be trusted and accurate. (iii) The SFC encapsulation which coveys important information about the SFC data-plane must be authenticated and/or encrypted. The exchange of SFC encapsulation information must be done via a trusted protocol and source. (iv) An adequate protocol providing isolation of different tenants in a multi-tenant cloud network should be in place.

Other challenges that are part of SFC include limited visibility in the end-to-end service delivery, due to multiple data-centers and administrative domains. The traffic steering in SFC can be unidirectional or bi-directional depending upon the requirement. The SF-like IDS and stateful firewall often require traffic to be steered in both forward and reverse direction. This can be quite challenging considering the network topology and constrained resource availability.

9.3 SDN- and NFV-based SFC

There have been some attempts at developing a standard for SFC deployments. The IETF is actively developing SFC architecture to perform flow classification for routing traffic between service functions.

The European Telecommunications Standards Institute (ETSI) utilizes a service architecture based on forwarding graphs to route traffic between VNFs using network service header. NFV allows the creation of a software version of service functions, e.g., virtual firewall, virtual routers, etc. By decoupling the VNFs from the physical devices on which they are being run, NFV reduces Capital Expenditures (CAPEX) and Operational Expenditures (OPEX). In terms of service deployment, this leads to increased agility and better response time [192]. NFV can provide dynamic policy enforcement and elastic resource flexing in a network.

The issues discussed in Table 9.1 can be solved by NFV-based solutions as discussed in Table 9.3. In the coming sections we will discuss detailed examples and real-world implementations of NFV technologies that help in dealing with SFC and middlebox deployment issues.

An example of VNF deployment over NFVI is segmented routing. Each *segment* is an ordered list of instructions that help in steering traffic through a specified path in SFC as discussed in [17]. In IPv6 architecture, the Segment Routing Header (SRH) carries a list of segments. The segment list consists of location information and traffic steering instructions.

Figure 9.4 shows end-to-end orchestrator that interacts with NFV managers for VNF configurations in SFC. The original IPv6 packet is appended inside the outer IPv6 packet along with SRH in this architecture. The incoming traffic is classified by edge-routers, which associate the traffic to their respective VNF chain. SRH consists of original packet and VNF chain information. The packet arriving at NFV node is processed using VNF connector, so that packet is redirected to the appropriate VNF.

OpenNFV [206] makes use of OpenStack based VNF Forwarding Graph (VNFFG) and OpenFlow-based SFC to implement SFC. The solution supports end-to-end packet delivery in a multi-tenant cloud network

TABLE 9.3

NFV Solution for Middlebox Failures

Failure Cause	NFV Solution
Misconfiguration	Centralized configuration policy checking using Openflow protocol
Overloading	Dynamic resource flexing and load balancing
Physical/Electric	Virtual Network Functions (VNF) replacing physical devices

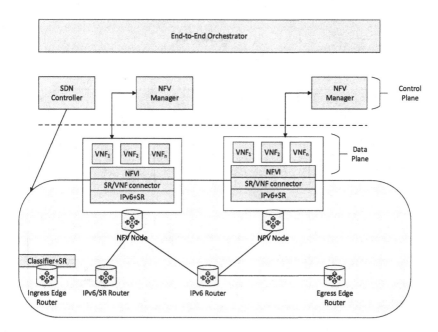

FIGURE 9.4
SFC segment routing using NFVI.

using ODL controller, which allows traffic to be steered through intermediate VNFs.

9.3.1 SDN as an Enabler of SFC

SDN allows separation of network traffic into data plane and control plane traffic using OpenFlow protocol. SDN has emerged as a strong candidate to serve as a de-facto protocol for deployment of VNFs in a cloud network. SDN provides programming abstractions required by SFC for traffic engineering and dynamic topology control. The capacity of SDN to dynamically manage VNF connections and the underlying data plane flows makes it an enabler for traffic steering capabilities required in SFC [268].

The control plane of SDN is responsible for management of SF instance, mapping of SF to a specific Service Function Path (SFP), installation of flow rules in Service Function Forwarding (SFF) devices (OpenFlow switches) [187]. SDN control plane interacts with SDN data plane elements via protocols such as OpenFlow. Figure 9.5 shows various interfaces that are used by SDN control plane to interact with SFC data plane elements that we discussed in previous sections.

SFC classifier which is responsible for traffic classification interacts with SFC control plane via interface C1 in Figure 9.5. The SFF devices such as

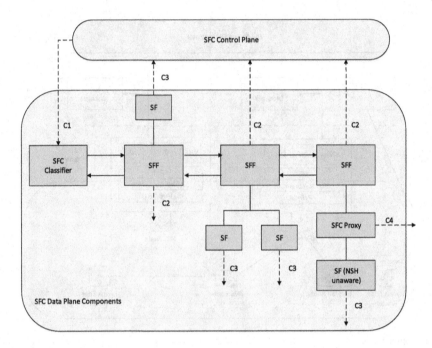

FIGURE 9.5
SDN enabled SFC architecture.

OVS report connectivity status of SFs attached to them to the control plane (interface C2) via OpenFlow protocol. Interface C3 is used by control plane to collect packet statistics and metadata from the NSH-aware SFs. The NSH-unaware SFs make use of SFC proxy shown in Figure 9.5 for a collection of statistics such as traffic workload, latency, etc. This information is relayed to the SDN controller over interface C4.

9.4 SFC Implementations

9.4.1 T-Nova: SDN-NFV-based SFC

T-Nova [163] is an SDN-based MANO framework. T-Nova is built using Opendaylight SDN controller and OpenStack cloud as key technologies. Most of the existing MANO deployments are based on the static configuration for the nodes with VNFs. This leads to inconsistencies between the workload requirements and features offered by the VNF node. Another key problem in current NFVI is the absence of monitoring capability that covers not only NFVI but also VNF services.

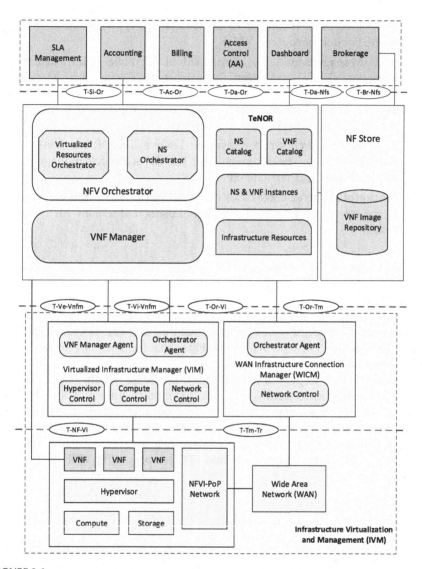

FIGURE 9.6
T-Nova architecture.

T-Nova automates the resource discovery, service mapping, service deployment and monitoring. T-Nova showcases the performance gains obtained using hardware acceleration. T-Nova architecture shown in Figure 9.6 is based on the European Telecommunications Standards Institute (ETSI) NFV ISG model with some added features. The key features in T-Nova fully modular Service Oriented Architecture (SOA) are as follows:

- **NFVI** layer, which consists of network services, physical and virtual layers (switches, routers, VMs).

- **NFV Management** layer, composed of Virtualized Infrastructure Manager (VIM) and Wan Infrastructure Connection Manager (WICM). The control plane of T-Nova utilizes OpenDaylight [188] SDN controller for network management and OpenStack [242] cloud for data-center-based compute resources.

- **TeNOR** is the orchestration layer for T-Nova. TeNOR is a store of all published VNFs. The design goal for TeNOR is lifecycle management of distributed and virtualized NFVI.

- **Marketplace** layer consists of all business functions and customer-facing interfaces, which allow external users to interact with T-Nova system.

T-Nova testbed has led to many SDN-based development efforts. SDK4SDN [128], aka NETwork FLOws for Cloud (Netfloc), is an opensource SDK for data center network programming development. Netfloc has integration support for SDN controller OpenDaylight. Netfloc provides Java interfaces and REST-based APIs for network programmability and end-to-end OpenStack-based data-centric network with OpenFlow enabled switches for leveraging SDN-based MANO.

SDK4SDN was deployed using Lithium release of the OpenDaylight controller. It was successfully tested on OpenStack Juno and Kilo releases. Netfloc showcases several important use case applications such as tenant isolation using lightweight non-GRE/VxLAN tunneling mechanism, layer 2 resilience using redundant datapath abstraction for link recovery and virtual switch reconfiguration and SFC for providing VNF traffic steering support.

9.4.2 Tacker: OpenStack-based SFC

Tacker [263] is an OpenStack-based VNF MANO platform. Tacker is based on ETSI MANO architecture and provides a functional stack to orchestrate end-to-end network services across VNFs.

The key components of Tacker are the following:

- **NFV Catalog** consists of VNF Descriptors, Network Service Descriptors and VNF Forwarding Graph Descriptors as shown in Figure 9.7.

- **VNFM** (VNF Manager) is responsible for basic lifecycle management operations such as create/update/delete, platform aware NFV load optimization, health monitoring, auto-scaling and VNF configuration management operations.

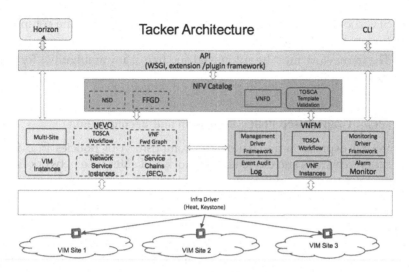

FIGURE 9.7
Tacker openstack NFV orchestration architecture.

- **NFVO**: NFV Orchestrator is responsible for VIM resource check and allocation, SFC management using VNF Forwarding Graph descriptor, VNF placement policy, network service deployment using decomposed VNFs.

Tacker initiative is an attempt to standardize the approach of SFC implementation using OpenStack and SDN Opendaylight controller.

9.5 Policy-Aware SFC

Security and traffic steering policies dictate how the traffic traversed between two endpoints in a cloud network. Policies help ensure optimal performance, redundancy, authentication and data integrity. Ensuring fulfillment of these objectives in SFC is a complex issue. There are several actors responsible for the formulation of these policies such as Application Service Provider (ASP), Internet Service Provider (ISP), Telecom Service Provider (TSP), etc. To further complicate the matter, policies can be static or dynamic (created on the fly). In this section, we analyze the efforts in the direction of efficient policy formulation and compliance in the context of SFC.

Some key requirements service function chaining that are provided by PGA frameworks are as follows:

- Each policy writer should be able to specify service chain policies independently.

- SFC should allow eager policy composition, satisfying the composition intent of the individual policies.
- The framework must be automated and free of any ambiguities for the network traffic.

9.5.1 PGA: Graph-based Policy Expression and Reconciliation

The natural expression and automatic composition of policies in different application scenarios such as tenant network and enterprise network is manual and error-prone tasks.

The expression of high-level policies, such as DNS policies and Firewall policies, into low-level network configuration requires coordination between admins and users, to manually check conflicts in the policy set.

The new network infrastructures such as SDN and NFV have multiple entities generating policies. Therefore, a mechanism that detects and resolves conflicts between policies is highly desired. Policy Graph Abstraction (PGA) [221] solves the problem of automatic expression, conflict-free and fast composition of network policies. PGA allows network policies to be represented as graph structures.

The users/admins/tenants can independently compose policies and submit them through Graph Composer through PGA User Interface (UI) shown in Figure 9.8. A policy specified at a high level can be composed to low-level configuration rules in this framework.

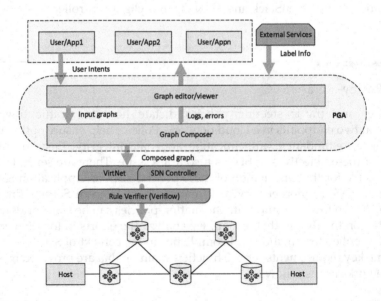

FIGURE 9.8
PGA system architecture.

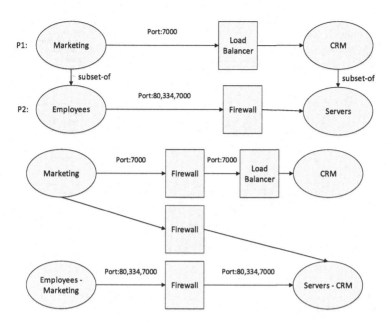

FIGURE 9.9
PGA policy composition example.

9.5.1.1 Policy Composition Example

Consider two policies - P1 and P2. Policy P1 says that only marketing employees can send traffic to the CRM server over port 7000. The traffic should go through the load balancer (LB). According to policy P2, employees should only access the servers installed in the company through ports 80, 3000 and 7000.

We can see in Figure 9.9 that policy P1 is completely encompassed by policy P2. If we use simple priority assignment measure and prioritize policy P1 over P2 since non-marketing employees will also be able to send traffic to the CRM server in that case. A correct way of policy composition as per PGA mechanism is shown in the figure. The marketing employees will need to go through Firewall for accessing any server which is not CRM. All employees who are not part of marketing will need to go through firewall SF. This arrangement may be able to achieve the desired service delivery objectives but will lead to performance overhead, since Firewall SF has been used thrice in the policy graph composition.

9.5.2 Group-based Policy

Group-based policy (GBP) is an automated intent-driven mechanism for mapping the subscriber traffic to the SFs. The automation of intent is achieved

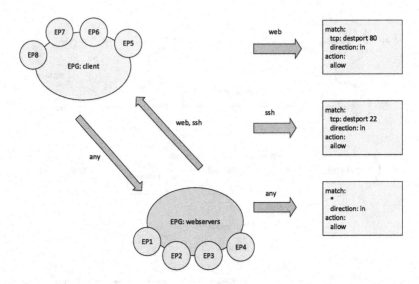

FIGURE 9.10
Endpoints, EPGs and contracts in GBP.

by mapping between the intent expressed and renderers that provide the capability to satisfy the intent. The entity that wants to communicate is known as *endpoint*, e.g., Container, Network Port. The endpoints that consist of common attributes can be composed into a *endpoint group (EPG)*. The EPGs can enter into *contracts* that determine the scope of communications that are allowed in GBP architecture. The EPGs can provide or consume contracts which determine the capabilities and requirements of various EPGs.

For instance, EPG: WebServers in Figure 9.10 are providing the contracts ssh and web. The *EPG: Client* can subscribe to the services provided as part of the web, ssh contracts. The *EPG: Client* is also providing the contract *any* that is subscribed by *EPG: WebServers*. As can be seen from the figure, the traffic with destination ports 22, 80 (web, ssh) is allowed from *EPG: WebServer* to *EPG: Client*, whereas for the opposite direction all traffic is allowed.

9.6 Secure Service Function Chaining

The order in which SFs are placed, the reliance on the underlay and overlay service delivery mechanisms, can lead to security issues and sub-optimal packet delivery in SFC. We discuss a few ordering and placement strategies and highlight the shortcomings below. Our SFC has the following requirements:

1. *Traffic coming into the network should be classified into different categories based on source IP address using Classifier SF.*

2. *Any traffic not part of data network security domain should be processed via Intrusion Detection System.*

3. *Data network traffic and SDN controller traffic should go through Load balancing SF.*

4. *Control plane traffic from SDN controller should be encrypted using public key encryption scheme.*

The nodes in an SFC architecture can be SFs or *Service Function Forwarders (SFFs)*. An SFF is responsible for forwarding traffic packets or frames received from a particular network segment to associated SFs using the information encapsulated in the packet. In Figure 9.11 the OVS bridge acts as SFF. The OVS bridges can communicate with each other using a tunnel network *tun0* shown in the figure. The SDN controller can communicate with OVS switches using the OpenFlow protocol. The SFs are connected to OVS bridges using tap interfaces *veth*. The traffic between OVS and controller is *Control Plane* traffic, whereas the traffic between OVS and SFs is *Dataplane* traffic.

The *Service Function Path (SFP)* is the actual path traversed by the packet/frame from source to destination in SFC after application of granular policies and operational constraints in SFC. For instance in Figure 9.11, there are three SFPs, i.e., SFP_1, SFP_2, SFP_3, corresponding to Data Net, Public Net and SDN controller traffic.

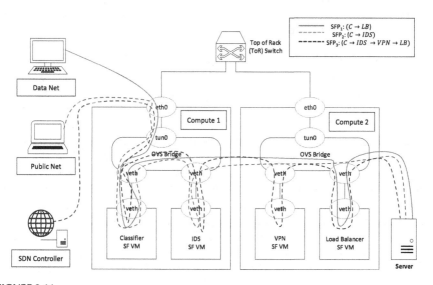

FIGURE 9.11
Service function chaining example in cloud network.

Strategy 1 *Order:* $C \rightarrow VPN \rightarrow IDS \rightarrow LB$.
Issue: SDN controller traffic needs to go through both VPN and IDS as per policy, placing VPN first (incorrect order) violates security objective. Thus, IDS should precede VPN, since VPN encrypts the traffic and IDS can operate only on the raw traffic.

Strategy 2 *Order:* $C \rightarrow LB \rightarrow IDS \rightarrow VPN$.
Issue: The traffic from SDN controller and data network has to go through classifier and load balancer. Malicious traffic could have been filtered out using IDS policy resulting in less impact on QoS offered by the load balancer. The incorrect placement leads to an efficiency issue. In order to preserve both security and efficiency constraints, we design better placement and ordering as shown in Figure 9.11 – *Strategy 3*.

Strategy 3 *Order:* $C \rightarrow IDS \rightarrow VPN \rightarrow LB$.
Efficient Placement is obtained in this strategy since unwanted traffic is filtered at IDS and load balancer has to deal with only legitimate traffic from *Data-Net* and *SDN Controller*.

Correct Ordering is obtained for SDN controller traffic. The traffic is passed in raw (un-encrypted) format through IDS, and later through VPN thus IDS has complete visibility.

9.6.1 Secure In Cloud Chaining

The security is a key requirement in an end-to-end service delivery. The outsourcing of network functions should satisfy confidentiality and privacy properties. Current secure SF outsourcing schemes incur performance overhead or are out-of-band (offline). Secure In Cloud Chaining (SICS) [277] framework encrypts each packet header based on in-cloud rule matching and incurs minimal performance overhead.

The SICS system in Figure 9.12 consists of a gateway for tunneling ingress and egress traffic to the cloud middleboxes. Incoming traffic headers are encrypted using AES encryption to ensure that cloud users are not able to access incoming traffic. The traffic is then passed through different SFs and decrypted at the enterprise gateway. The decrypted traffic is sent to the internal network. Each packet is assigned a label to ensure appropriate SFP is followed and rule matching is successful when the packet traverses through a sequence of middleboxes. The same operations are applied to the traffic going from the internal network to the internet in reverse order. The traffic is encrypted at the enterprise gateway and sent to the internet.

The label-based approach provides efficient packet lookup. A 16-bit label is able to represent one million 5-tuple rules. The lookup overhead for the matched labels using hashtable is $O(1)$. A label can only reveal information about the actions and forwarding behavior of the middleboxes, thus preserves privacy.

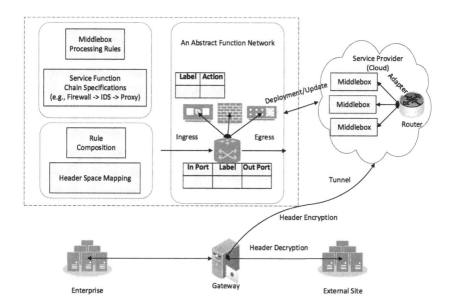

FIGURE 9.12
SICS design framework.

9.6.2 SFC Using Network Security Defense Patterns

In a multi-tenant cloud network, each tenant can have different security and service provisioning requirements. SDN with NFV can help inflexible composition of network functions. Most VNF deployment solutions consider only the performance aspect of the SFC. The security needs are often loosely coupled with other SFC objectives. For instance *Strategy 1* in Figure 9.11 considers placement of VPN service function before IDS, which violates security objective, whereas *Strategy 3* provides correct ordering of SFs. Similarly, other security requirements such as DDoS blocking close to the source should be considered as a part of SFC deployment.

The high-level security needs can be mapped to security patterns and the security constraints can be mapped to appropriate SFC deployment constraints as discussed in Network Security Defense Pattern-based SFC framework [244] as can be seen in Figure 9.13. Table 9.4 highlights various NSDP that can help in achieving security and ordering objectives in SFC.

The NSDPs represent security patterns which can be combined with deployment constraints in order to achieve SFC. The deployment constraints can be broadly classified into *Region, Co-location, Distribution, and Waypoint*.

- **Region** constraint forces the SF to be placed in a specific region or zone. For instance, two different IDSs can be placed close to each other (same region) in order to achieve collaboration.

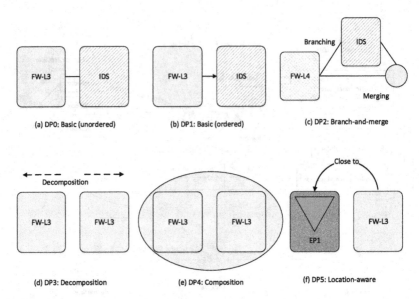

FIGURE 9.13
Network security defense patterns in SFC.

TABLE 9.4

NSDP-Aware SFC

NSDP	Description
Basic Unordered	NSDP in which packet traversal through SFs does not need to follow any strict ordering, e.g., Layer 3 (L3) Firewall and IDS – Figure 9.13(a).
Basic Ordered	NSDP where one SF needs to be traversed before another in order to meet security or efficiency objectives, e.g., IDS placement before VPN – Figure 9.13(b).
Branch-and-Merge	SFC where incoming traffic is required to be split into two branches in order to pass through different service function paths – Figure 9.13(c).
Decomposition	NSDP which requires splitting of the functionality of a given SF into two different subsets of the same type for the purpose of de-localization. For instance an implementation of distributed firewall via splitting of firewall functionality – Figure 9.13(d).
Composition	NSDP can sometimes require the merging of the functionality of two SFs in order to achieve some desired properties such as state-aware SFC. For example, a stateful firewall can have two bi-directional traffic SFs merged together – Figure 9.13(e).
Probe Point	SFC can require placement of an SF at a certain distance from a location, e.g., placement of L3 firewall close to the endpoint and DDoS SF close to the location of the victim – Figure 9.13(f).

- **Co-Location** constraint requires placement of several similar or dissimilar SFs on the same node. Two security functions with traffic flow in opposite directions can be placed on the same node in order to achieve a stateful SFC property.

- **Distribution** requires placement of SFs on different nodes. A distributed intrusion detection mechanism can require placement of IDS on different nodes in order to achieve fast and complete security coverage of the entire network.

- **Waypoint** constraint requires traversal of traffic flows from different nodes through a given SF. If we need DPI for suspicious traffic pattern, we may require such traffic to pass through DPI SF.

Figure 9.14 shows three NSDPs, i.e., FW L3, IPS and WAF that are present between EP1 and EP2 (Web Server). The NSDP basic location-aware, composition and decomposition are incorporated in this SFC deployment. We describe the NSDP in this use case scenario below:

- **Decomposition and Location Awareness**: The layer 3 firewall should be decomposed and placed close to source and destination. The L3-FW is decomposed into FW L3_1 and FW L3_2, so that one firewall can be placed near EP1 and another near the destination node EP2. This deployment also satisfies the location awareness criteria.

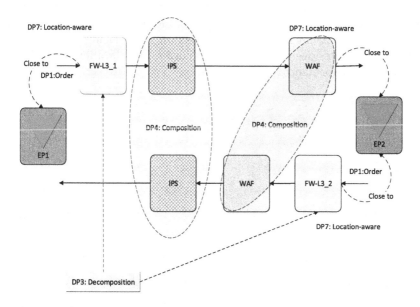

FIGURE 9.14
SFC with NSDP composition, decomposition and location awareness.

- **Composition**: We want the traffic to be steered through the IDS and Web Application Firewall (WAF) while traveling from source to destination. Thus, we use a combination function for combining the functionality of WAF and IDS into one SF for SFC deployment.

Summary

Service Function Chaining (SFC) architecture allows chaining of various network functions such as Firewall, Load Balancer, and Intrusion Detection System (IDS) in order to steer traffic between two network endpoints. Challenges in SFC such as topological dependencies, ordering, transport protocol dependencies, traffic selection, policy routing, and security constraints imposed by various underlay and overlay technologies limit the service delivery and possible deployment options in a network.

SDN and NFV-based SFC help in dealing with challenges in SFC. The NFV helps in virtualization of various hardware-based SFs, thus reducing CAPEX and OPEX. The use of SDN allows programmability in the network, thus network monitoring, route optimization, SF ordering and traffic engineering are possible by using SDN architecture. SFC implementation frameworks such as T-Nova, Netfloc, and Tacker serve as a guideline for the adoption of SFC in a cloud network. Policy compliance is critical in a cloud network. Policies can be composed into a policy graph in order to identify the order and dependencies between various SFs in the network. Secured Service Function Chaining (SSFC) can be achieved by identifying requirements such as traffic encryption, load-balancing, intrusion detection, and placing the VNFs in an order that achieves desired end-to-end security while limiting the impact on the network performance.

10

Security Policy Management in Distributed SDN Environments

Security policy management can be thought of as the framework for specification of an authentication and authorization policy, and the translation of this policy into information that can be used by devices to control access, management of key distribution, audit of security activities and information leakage [254]. This authorization usually pertains to permitting or denying access to resources or information [196]. Security management almost always also includes actions to be taken if any violations are detected.

Given the rapid growth in the scale of networks being deployed, traditional methods which rely on trained personnel to implement and manage information security has become more time consuming, and error-prone [274]. Maintaining a mandated network security scheme for large-scale data center networks and distributed environments is a formidable challenge. It is to this end that several policy-driven management techniques have been suggested [107]. By separating out security policies from their low-level implementation and enforcement in the network, such methodologies simplify network management while paving the way for seamless growth of the network [177,247].

The ease of programmability in SDN makes it a great platform for implementation of various initiatives that involve application deployment, dynamic topology changes, and decentralized network management in a multi-tenant data center environment. This programmability gives great flexibility in implementing applications as well as security solutions. Having flexibility also lends to the possibility of policy conflicts - both intentional and unintended. Throw in the ability to have multiple SDN controllers to complicate the situation by having potentially different policies in place for the same traffic! Thus, implementing security solutions in such an environment is fraught with policy conflicts and consistency issues with the hardness of this problem being affected by the distribution scheme for the SDN controllers.

In this chapter, the implications of security policy conflicts is discussed first. Next a formalism for flow rule conflicts in SDN environments is described. A comprehensive conflict detection and resolution models ensure that no two flow rules in a distributed SDN-based cloud environment have conflicts at any layer; thereby assuring consistent conflict-free security policy implementation and preventing information leakage. Strategies for prioritizing and unassisted resolution of these conflicts are also detailed.

10.1 Background

The definition of policy itself is rather ambiguous and is often something of debate. Policies could be thought of as a specific way to dynamically implement static requirements [95]. Separating out the policy from the requirement enables them to be altered and adjusted to the environment or modified to improve performance, all the while ensuring that they still adhere to the requirement. In fact, the requirement can be used as a gauge to verify the functionality of the policy. A policy hierarchy that represents the relationships between different levels of policy abstractions, as shown in Figure 10.1, is generally accepted to be [54,75,193,194]:

- Requirements, high-level abstract policies or management goals: These are generally natural language statements such as Service Level Agreements (SLA) or business goals. They are usually not enforceable at a device level, and are implemented using a lower abstraction level.

- Specification-level or network-level policies: These are specified by a human administrator in a precise format to provide abstractions for device-level implementations. These policies must be specific enough to drive automation.

- Low-level policies or device configurations: These are implemented on the devices themselves. These are often the bottleneck to both scalability, performance and interoperability.

The level of policy most relevant to our study is specification-level policy, or network policy. We adopt the definition of a network policy from the work of Damianou [75].

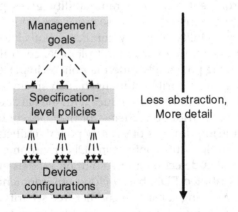

FIGURE 10.1
Policy hierarchy.

Definition 1 *A network policy consists of rules which define relationships between network resources and the network elements that provide those resources. Network policies manage and control the accessibility, reliability and the QoS experienced by networked applications and users.*

The IETF policy model [260] specifies that network policies be considered as rules that specify actions to be taken when certain conditions are met, described by the syntax below.

```
IF <condition (s)> THEN <action (s)>
```

While the syntax described follows the Condition-Action paradigm of most Policy Core Information Model (PCIM) [196] rules, flow rules written in this syntax follow similar semantics to an obligation [75] in the form of an Event-Condition-Action (ECA) paradigm [35] from event-driven architectures, with an implicit event trigger in case of a match. To help determine the action set when multiple conditions are met, most policies are associated with a priority value. Alternately, instead of specifying an explicit priority, a role-based priority may be assigned to the policies depending on the origination point of the policy.

A typical policy-based management architecture as per the PCIM is shown in Figure 10.2. A Policy Manager (PM) serves to facilitate policy formulation, analysis and verification. Once verified, the policies are stored in a Policy Repository (PR). The Policy Decision Point (PDP) actively monitors the system

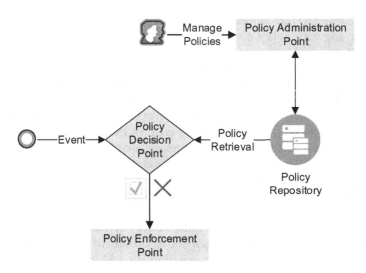

FIGURE 10.2
Policy management framework.

for specific events. When triggered by certain conditions, the Policy Enforcement Point (PEP) comes into play to enforce the policy actions.

Several issues with regard to policy definition and implementation, such as policy storage and enforcement in a distributed environment are addressed natively in SDN, wherein the SDN controller can act as a PR and PDP, while the SDN switches can act as the PEP.

10.2 Related Work

10.2.1 Firewall Rule Conflicts

There have been several attempts to classify policy conflicts. Lupu and Sloman [175,176] describe conflict analysis for management policies, using a tool to conduct offline detection of conflicts in a large-scale distributed system. Eppstein and Muthukrishnan [86] use a deal with the packet classification and filter conflict detection problem. They use KD tree [44] to verify if two rules apply different actions on the same packet. This misses out on some conflict classification types that involve sub-optimal rules. Fu et al. [95] manage policies as they apply to IPSec tunnels in both inter- and intra-domain environments. Hari et al. [108] present a conflict detection and resolution algorithm using a k-tuple filter that grows linearly. However, the seminal work by Al-Shaer and Hamed [23] is often used to classify firewall conflicts. In that work, the authors also introduce the Firewall Policy Editor to provide a simple representation allowing for easy human recognition of firewall rule conflicts. The authors extend this work into a distributed environment in Firewall Policy Advisor (FPA) [22], which identifies where to insert new firewall rules to not incur any new policy conflicts. In FPA, the authors also introduce basic visualization by displaying simplified versions of complex firewall rules, and show firewall rule conflicts in tabular format.

Firmato [38] is a firewall management toolkit that helps users separate out security policy and the underlying network topology. It uses a Model Definition Language (MDL) to translate the security model into configuration files for Lucent managed firewalls. Firmato ensures there are no policy conflicts in the system and that the rules in the firewall are up to date with the security policy. Fireman [293] detects misconfiguration stemming from: *a*) violations of user-specified security policies; and *b*) inconsistencies among firewall rules. It parses firewall configurations and converts them into an operational semantics representation, which is then sent to the administrator for decision making. Unlike FPA, Fireman also recognizes inter-firewall rule conflicts.

FAME [120] is a conflict management environment to detect and resolve conflicts by using rule-based segmentation. In FAME, the authors used a matrix to represent conflicting and non-conflicting address segments; but failed while trying to represent larger rule sets.

Capretta et al. [55] proposed a formalization of conflict detection for firewalls, but constrained themselves to only look at rules where the actions are different; thereby missing out on some conflict classes. Rei [141] is a language based on deontic logic [281] that defines security policies as possible actions on a resource. All policies in Rei are free of conflicts due to the presence of meta-policies defined by an administrator, which are used to resolve conflicts. If a meta-policy that covers the conflict does not exist, by default the deny action is prioritized. FLIP [295] is a high-level firewall configuration policy language in which security policies are translated into lower level configuration to be loaded onto the devices. Since FLIP is a centralized configuration generation point, the rules generated will be conflict-free due to FLIP preventing overlap of any kind [220].

Fang [182] is a tool that reads in the vendor specific configuration files and converts them into an internal representation, which is then presented to the administrator in a tabular form in simple text. While it is one of the earliest work in visualization of rule conflicts, it is devoid of any graphics. While it is a step to making security configurations vendor agnostic, it does not display any relation between conflicting rules. The onus is on an experienced administrator to submit the right query that would present the conflict. Policy-Vis [269] used overlapping bars to represent conflict types, and colors to represent the action. However, the conflicts are visible only when a certain scope is defined. A sunburst visualization is used by Mansmann et al. [179] to visualize the rule set, but does not provide any visualization for flow rule conflicts. None of the above works provide scalable rule conflict visualization on-demand to the administrator, with high-level conflict categorization and granular information.

10.2.2 SDN Security and SDN Policy Management

While advances in SDN have made it central to deployment of a cloud environment, security mechanisms in SDN trail its applications. A basic SDN firewall was introduced as part of Floodlight [5], wherein the first packet in a new flow is sent to the controller to be matched against a set of flow rules. The resulting action set is then sent to the OpenFlow switch. The action set is applied for the current flow, and cached for enforcement on all future flows matching the same conditions. In case of a dynamic security policy update to the controller, the OpenFlow switches are oblivious to this situation and could implement a dated action set. Javid et al. [135] built a layer-2 firewall for an SDN-based cloud environment using a tree topology for a small network using a POX controller and restricted traffic flow as desired. Suh et al. [261] illustrated a proof-of-concept version of a traditional layer-3 firewall over an SDN controller. An application layer firewall using SDN was demonstrated by Shieha [250].

FRESCO [251] allows for the implementation of security services in an OpenFlow environment by providing reusable modules accessible through a Python API. To address conflicts that might arise in an OpenFlow environment, FRESCO introduces a Security Enforcement Kernel (SEK) that prioritizes rules to assist in conflict resolution, but does not tackle complex flow rule inconsistencies. FortNOX [218] is an extension to the NOX controller that implements role-based and signature-based enforcement to ensure applications running on the controller do not circumvent the existing security policy, thereby enforcing policy compliance. In FortNOX, reusable modules are used to protect the flow installation mechanism against adversaries, but conflict analysis is conducted only between a new flow rule and existing rules, without considering dependencies within flow tables. Thus, implementing FortNOX in a distributed environment would be challenging. Decision making in FortNOX seems to follow a least permissive strategy instead of making a decision keeping the holistic nature of the environment in mind. Moreover, it uses only layer-3 and layer-4 information for conflict detection, which we believe is incomplete since SDN flow rules could use purely layer-2 addresses for decision making. In addition, FortNOX would not be able to handle partial flow rule conflicts.

In their work, Cholvy and Cuppens [60] conduct consistency analysis of security policies, focusing on access control policy. Much of their study focuses on defining and ensuring security policy consistency based on deontic logic, like Rei. But paradoxes do exist in deontic logic [84] and hence such work would be beneficial only in a complementary manner to any security implementation involving network devices such as firewalls.

Natarajan et al. [202] study two different conflict detection techniques for flow rules. The first approach internally represents flow rules using a combination radix trie and a hash, which are then used to identify conflicts. The second approach is an ontology-based detection system. However, the authors do not discuss conflicts in distributed environments, or how to resolve any conflicts.

FlowChecker [21] identifies intra-switch conflicts within a single flow table using Binary Decision Diagrams (BDD) [109]. Certain conflicts in SDN networks can also be determined by expanding the work of Gu et al. [102] in detecting anomalies in network traffic tailored to a SDN environment. However, these would be limited to detecting network invariants that can be detected by comparing current network traffic with a baseline distribution.

Pyretic [195], a high-level language written in Python courtesy of the Frenetic project [92], allows users to write modular applications. Modularization ensures that rules installed to perform one task do not override other rules. Using a mathematical modeling approach to packet processing, Pyretic compares the list of rules as functions that use a packet as an input, and have a set of zero or more packets as output. Given its mathematical base, Pyretic deals effectively with direct policy conflicts, by placing them in a prioritized rule set much like the OpenFlow flow table. However, indirect security violations or inconsistencies in a distributed SDN environment cannot be handled by Pyretic without a flow tracking mechanism such as the one discussed by Fayazbakhsh et al. [89].

VeriFlow [152] is a proposed layer between the controller and switches which conducts real time verification of rules being inserted. It uses search rules based on Equivalence Classes (ECs) to maintain relationships and determine which policies would be affected in case of a change. Thus, it can verify that flow rules being implemented have no errors due to their dependence on faulty switch firmware, control plane communication, reachability issues, configuration updates on the network, routing loops, etc. Like VeriFlow, Net-Plumber [148] sits between the controller and switches. Using header space analysis [149], it ensures that any update to a policy is compared to all dependent policies to prevent and report violations.

FlowGuard [121] is a security tool specifically designed to resolve security policy violations in an OpenFlow network. FlowGuard examines incoming policy updates and determines flow violations in addition to performing stateful monitoring. It uses several strategies to refine policies, most of which include rejecting a violating flow.

Research in SDN security enforcement such as AVANT-GUARD [252] allow for development of security enforcement kernels, and threat detection to applications. In Sphinx [80], the authors extend attack detection in SDN to include a broader class of attacks including untrusted switches and hosts. However, these security solutions implicitly assume the presence of a conflict-free security policy for implementation, and do not address the problem of conflicting flow rules.

Since a flow can be defined using addresses in multiple layers, policy checking approaches in SDN should differ from traditional approaches by being able to consider indirect security violations, partial violations or cross-layer conflicts. However, none of the works discussed above tackle these problems. Moreover, they appear not to fully leverage the SDN paradigm that lets flow rules do traffic shaping in addition to implementing accept/deny security policy. To that end, we propose Brew, a framework that considers cross-layer dependencies while ensuring conflict-free policies in a distributed SDN-based environment. Additionally, Brew analyzes traffic shaping policies including rate limiting policies along with security policies to detect and resolve direct, indirect and partial conflicts.

The remainder of this chapter details flow rules, flow rule conflicts and resolution mechanisms.

10.3 Flow Rules

OpenFlow v1.3.1 specifications [14] describe a flow rule to consist of the following fields:

- Priority which describes the precedence of the rule, and is defined in the range [1, 65535]. Higher priority values are preferred over lower values. If left unspecified, the priority field defaults to 32768.

- Match fields which consist of protocol specific header information, hardware addresses, and metadata that is used to match incoming flows. In all, the basic class in OpenFlow 1.3.1 can match thirty-nine different values, amongst which thirteen values are required to be handled by all switches. These are: *a*) ingress port; *b*) Ethernet source address; *c*) Ethernet destination address; *d*) Ethernet type; *e*) IPv4 source address; *f*) IPv4 destination address; *g*) IPv6 source address; *h*) IPv6 destination address; *i*) IPv4 or IPv6 protocol number; *j*) TCP source port; *k*) TCP destination port; *l*) UDP source port; and *m*) UDP destination port.

- Packet counters which keep track of the number of packets that utilize the flow rule, and are updated each time a packet match is detected. About forty different counters are specified in the OpenFlow 1.3.1 specification.

- An action set that contains instructions on what to do when a matching flow is detected. Associated actions can: *a*) forward packets through a specified port; *b*) flood the packet on all ports; *c*) change QoS; *d*) encapsulate; *e*) encrypt; *f*) rate limit; *g*) drop the packet; and *h*) be customized using various `set-field` actions. The action sets are carried between flow tables, in cases where pipeline processing of flow tables is in effect. The complete list of match fields is shown in Table 10.1.

- Timeouts which specify the maximum amount of time or idle time before a switch would consider the flow rule expired.

- A cookie value chosen by the controller. This value is not visible to the switches, and therefore not used when processing packets. It may however be used by the controller to filter flow statistics, flow modification and flow deletion.

Since packet counters, timeouts and cookie values are not central to handling flow rule conflicts, in the remainder of this book, we limit discussion of flow rules to priority, match fields and action set fields. Table 10.2 shows a sample flow table rules with the selected fields present. The data in the table has been written to be human readable. The mapping of the columns is as follows: *a*) Rule #, present only to refer to the rules in this book and not present in OpenFlow; *b*) Priority; *c*) Source MAC, which is specified using the Ethernet source address field; *d*) Destination MAC, which is specified using the Ethernet destination address field; *e*) Source IP, which is specified using the IPv4 source address field; *f*) Destination IP, which is specified using the IPv4 destination address field; *g*) Protocol, which is specified using the IPv4 protocol field; *h*) Source Port, which is specified using either the TCP source field or the UDP source field; *i*) Destination Port, which is specified using either the

TABLE 10.1

Flow Table Actions

Argument	Description
OFPAT_OUTPUT	Output to switch port
OFPAT_COPY_TTL_OUT	Copy TTL "outwards" from next-to-outermost to outermost
OFPAT_COPY_TTL_IN	Copy TTL "inwards" from outermost to next-to-outermost
OFPAT_SET_MPLS_TTL	MPLS TTL
OFPAT_DEC_MPLS_TTL	Decrement MPLS TTL
OFPAT_PUSH_VLAN	Push a new VLAN tag
OFPAT_POP_VLAN	Pop the outer VLAN tag
OFPAT_PUSH_MPLS	Push a new MPLS tag
OFPAT_POP_MPLS	Pop the outer MPLS tag
OFPAT_SET_QUEUE	Set queue ID when outputting to a port
OFPAT_GROUP	Apply group
OFPAT_SET_NW_TTL	IP TTL
OFPAT_DEC_NW_TTL	Decrement IP TTL
OFPAT_SET_FIELD	Set a header field using OXM TLV format
OFPAT_PUSH_PBB	Push a new PBB service tag (I-TAG)
OFPAT_POP_PBB	Pop the outer PBB service tag (I-TAG)
OFPAT_EXPERIMENTER	Experimenter defined

TCP destination field or the UDP destination field; and *j*) Action, which is specified in the action set but simplified here to just forward and drop. All required fields that are ignored in Table 10.2 can be assumed to be wildcarded.

10.3.1 Security Policies Using Flow Rules

Due to the ability to alter headers from multiple layers of the OSI stack, flow rules in the OpenFlow protocol can inherently be used for traffic forwarding, routing and traffic shaping. Research has shown that, in addition to traffic manipulation functionalities, most security policies can be transferred into flow entries and deployed on OpenFlow devices [171].

While several security mechanisms implemented in traditional environments depend on routing traffic through middleboxes [140,243,248], it has been demonstrated that integrating processing into the network is just as effective [169]. The centralized control in the SDN paradigm can make this integration simple and elegant. Models to implement traditional security functions such as firewall rules, an IDS and Network Address Translation (NAT) rules in software have been demonstrated to be successful [27,98,213]. SIMPLE, a framework that achieves OpenFlow-based enforcement of middlebox policies, has been demonstrated in [222]. Contextual

TABLE 10.2

Flow Table Example

Rule #	Priority	Source MAC	Dest MAC	Source IP
1	51	*	*	10.5.5.0/24
2	50	*	*	10.5.5.5
3	52	*	*	10.5.5.5
4	53	*	*	10.5.5.0/24
5	54	*	*	10.5.5.5
6	51	*	*	10.5.5.0/16
7	55	*	*	10.5.5.5
8	57	11:11:11:11:11:ab	11:11:aa:aa:11:21	*
9	58	*	*	*

Rule #	Dest IP	Protocol	Dest Port	Action
1	10.1.1.63	tcp	*	forward
2	10.1.1.63	tcp	80	forward
3	10.1.1.0/24	tcp	*	forward
4	10.1.1.63	tcp	*	drop
5	10.1.1.63	tcp	*	drop
6	10.1.1.63	tcp	*	drop
7	10.1.1.0/24	tcp	10-17	drop
8	*	*	*	forward
9	*	tcp	80	drop

Note: All required fields not shown here are assumed to be wildcarded.

meaning to assist in the implementation of middlebox policies using Flow-Tags was demonstrated in [89]. Further, an OpenFlow-based multi-level security system that implements desired security policies using flow rules to accomplish network traffic monitoring as well as verification of packet contents has been successfully implemented [172]. François et al. [94] survey these and several other security implementations using OpenFlow.

Four of the most generic security-related policies are firewall, IPS/IDS, load balancing and NAT rules, each of which can be expressed using the flow rule tuple. A typical firewall rule that blocks all Telnet traffic can be specified in OpenFlow as follows. Note that nw_proto=6 signifies TCP.

```
priority = 51, nw_proto = 6, tp_dst = 23, actions = drop
```

Similarly, a load balancer policy, IPS/IDS policy or a NAT policy could be implemented by modifying the layer-3 source or destination address to send the flow to a specific device as follows:

```
priority = 51, nw_src = 10.5.5.5, nw_dst = 10.1.1.1,
actions = mod_nw_dst = 10.1.1.63, output :3
```

10.3.2 Flow Rule Model

In order to formally create a model that describes flow rules in an SDN-based cloud environment, an address n is defined in Definition 5.

Definition 2 *A frame space of a rule r is the subset of all possible 6-byte hexadecimal numbers representing OSI layer-2 (MAC) addresses, and is expressed as a 2-tuple (ϵ_s, ϵ_d) with subscript s denoting source and d denoting destination addresses.*

Definition 3 *A packet space of a rule r is the subset of all possible 32-bit numbers representing OSI layer-3 (IPv4) addresses, and is expressed as a 2-tuple (ζ_s, ζ_d) with subscript s denoting source and d denoting destination addresses.*

Definition 4 *A segment space of a rule r is the subset of all possible 16-bit numbers representing OSI layer-4 (TCP/UDP) addresses, and is expressed as a 2-tuple (η_s, η_d) with subscript s denoting source and d denoting destination addresses.*

Definition 5 *An address space n of a rule r is the 6-tuple representing the frame space, packet space and segment space, and is expressed as $(\epsilon_s, \epsilon_d, \zeta_s, \zeta_d, \eta_s, \eta_d)$, with subscript s denoting source and d denoting destination addresses. An address space is interchangeably called an address in this book.*

If N is the universal set of address spaces, we have:

Definition 6 *A flow rule r is a function $f : N \to N$ that transforms n to n', where n' is $(\epsilon'_s, \epsilon'_d, \zeta'_s, \zeta'_d, \eta'_s, \eta'_d)$ together with an associated action set a, that can have any of the values from Table 10.1. Thus,*

$$r := f(n) \rightsquigarrow a$$

The `set-field` capabilities in the action fields of the rules ensure that any, all or none of the fields in n may be modified as a result of the transform function f. Considering cases where the action set a is a pointer to a different flow table, we can apply the transform function on the result of the original transform function n'. Formally, if $r := f(n) \rightsquigarrow a$; $f(n) = n'$ and $a := g(n') \rightsquigarrow a'$ then,

$$r := g(f(n)) \rightsquigarrow a'$$

Thus, multiple rules applied in succession to the same input address space can simply be modeled as a composite function. It must be noted that the complexity of the flow rule composition function would be exponential in nature, since each flow rule could have multiple actions, each of which could recursively lead to multiple actions.

10.4 Flow Rule Management Challenges

Unlike traditional firewall rules, flow rules can match more than just OSI layer-3 and layer-4 headers making them inherently more complex by virtue of having additional variables to consider. Since wildcard rules are allowed in OpenFlow, a partial conflict[1] of a flow policy could occur, thereby adding complexity to the resolution of conflicting flow rules.

As discussed in Section 10.3, actions that can be applied on a match include forwarding to specific ports on the switch, flooding the packet, changing its QoS levels, dropping the packet, encapsulating, encrypting, rate limiting or even customizable actions using various set-field actions. The set-field functionality is a double-edged sword. One the one hand, it provides flexibility and allows the OpenFlow protocol to define complex virtual paths for traffic, and helps assert granular control. Cross-layer interaction is bolstered by virtue of having flow rules using set-field actions to change packet headers at several layers dynamically. But it also introduces significant management challenges, such as the origin binding problem [89,219].

OpenFlow specifications on how a flow match is determined are ambiguous, with the specifications stating that an incoming packet is compared against the match fields in priority order, and the first complete match is selected [14]. However, when the OFPFF_CHECK_OVERLAP flag is not set in the controller, multiple flow entries with the same priority can be set, in which case, the selected flow entry is explicitly undefined [14]. This is often the case in multi-tenant data centers, since setting the OFPFF_CHECK_ OVERLAP flag would result in capping the size of each flow table to 65,535 entries. When multiple matching rules with the same priority are encountered, directions on how to deal with the issue are unclear, and not standard across different implementations. While some implementations install *sensible* behavior such as more specific flows taking precedence over less specific flows, this is not specified in the OpenFlow specification [14], and not implemented in OVS. For instance, the only constraint OVS places requires flow descriptions to be in normal form, i.e., a flow can specify details for a particular layer header only if the protocol field in its lower layers is populated. That is, if the layer-2 protocol type dl_type is wildcarded, indicating use of any layer-3 protocol, then the flow rule cannot specify layer-3 IPv4 source and destination addresses. But, this requirement only does not prevent conflict causing scenarios. Furthermore, research has shown that despite there being clear prioritization rules in OpenFlow, certain hardware OpenFlow switches ignore priorities and treat rules installed later as more important [165]. Needless to say, ambiguity is highly undesirable in any security implementation, and preventing conflicts in flow rules is key.

---

[1] Caused when there is a partial overlap in the address spaces of the rules, as described in Section 10.3

Security implementations using SDN leverage the ability to make dynamic changes to the network and system configurations to have a lean, agile and secure environment. Since this usually results in environments that are constantly in flux, ensuring synchronization of the flow rules on all the distributed controllers is challenging. As and when the logical topology changes, the flow rules in place must be modified in accordance to ensure policy compliance. Additionally, ensuring that the changing flow rules are always in line with the security policy of the organization is not trivial [103].

Finally, flow rules in an SDN environment can be generated by any number of applications rather than just from an administrator. While this can reduce the workload on the administrator and help with chronic complexity management, there exists a potential for misplaced priorities between some of the flow rule generation points. Besides, an application acting maliciously can wreak havoc across the environment if not detected early enough [18]. Having multiple applications with the ability to concurrently update flow rules can lead to unexpected conditions if the holistic nature of the environment is not considered. For example, consider a load balancing and a DPI application running on the controller. If the DPI detects an intrusion on a node, it would attempt to migrate traffic off it. However, if the load balancer was responsible for allocating new incoming connections to the device with the fewest number of active connections, it might effectively sabotage the attempts of the DPI program.

To summarize, flow rule management is more complex than rule management in a traditional environment because:

- Match conditions cover more fields than in traditional environments.
- The set-field actions lead to cross-layer interaction in SDN flows.
- Flow rule priority field is not unique, and there is no standard on how to handle flow rules with the same priority.
- Ensuring synchronicity of rules in a multiple controller environment is not trivial when the topology is constantly changing.
- There are multiple generation points for flow rules, and there exist potential for some of the generation points to not have the same priorities as the administrator.

10.4.1 Motivating Scenarios

One of the major benefits of using SDN to implement a cloud environment is the ability to have multiple applications run on the SDN controller, each of which has complete knowledge of the cloud environment. This can be leveraged by the cloud provider to provide Security-as-a-Service (SaaS). A few potential examples of services in a SaaS suite are Firewalls, VPN, IDS, IPS, MTD, etc. Implementing a management system that only specifies security policies without tackling topological interaction amongst constituent members has always been a recipe for conflicts [95].

With the SDN controller having visibility into the entire system topology along with the policies being implemented, several of the conflict causing scenarios in traditional networks were handled. However, there are several instances where conflicts can creep into the flow table such as policy inconsistencies caused by: _a_) service chain processing where multiple flow tables that handle the same flow might have conflicting actions; _b_) VPN implementations that modify header content could result in flow rules inadvertently being applied to a certain flow; _c_) flow rule injection by different modules (using the northbound API provided by the controller) could have conflicting actions for the same flow; _d_) matching on different OSI layer addresses resulting in different actions; and _e_) administrator error. This list, while incomplete, goes to show how prevalent policy conflicts in SDN-based cloud environments could be.

Three distinct case studies in an SDN-based cloud environment where the security of the environment is put at risk due to flow rule conflicts are discussed next. The first scenario serves as an example where rules from different applications conflict with each other, and the second scenario serves as an example where rules from a single module might cause conflicts due to the dynamism in the environment. The last scenario once again discusses how inconsistent view of the network state results in different applications inserting flow rules with incomplete information.

10.4.1.1 Case Study 1: MTD

Traditional approaches to addressing security issues in a dynamic, distributed environment concerned themselves with implementing security through individual components, and not considering security holistically. This leads to two critical weaknesses: _a_) defense against insider attack is minimal; and thus _b_) when perimeter defenses fail, internal systems are ripe for the picking. As a counter, security applications that implement MTD is a topic that is hotly researched.

MTD techniques have been devised as a tactic wherein security of a cloud environment is enhanced by having a rapidly evolving system with a variable attack surface, giving defenders an information advantage [64]. An effective countermeasure used in MTD is network address switching [284], which can be accomplished in SDN with great ease. Since an MTD application could dynamically and rapidly inject new flow rules into an environment, it could lead to conflicts between the new and old flow rules.

In the data center network shown in Figure 10.3, we have Tenant A hosting a web farm. Being security conscious, only traffic on TCP port 443 is allowed into the IP addresses that belong to the web servers. When an attack directed against host _A_2 has been detected, the MTD application responds with countermeasures and takes two actions: _a_) a new web server (host _A_3) is spawned to handle the load of host _A_2; and _b_) the IP for host _A_2 is migrated to the Honeypot network and assigned to host _Z_1.

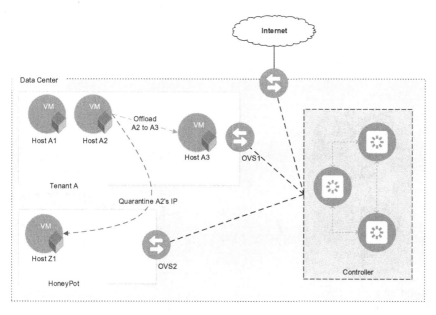

FIGURE 10.3
Policy conflicts in SDN-based cloud caused by MTD.

To run forensics, isolate and incapacitate the attacker, the Honeypot network permits all inbound traffic, but restricts egress traffic to other sections of the data center. These actions result in new flow rules being injected into the flow table that: *a*) permits *all* traffic inbound to the IP that originally belonged to host *A*2, but now belongs to host *Z*1; *b*) modifies an incoming packet's destination address from host *A*2 to host *A*3 if the source is considered to be a non-adversarial source; *c*) stops all outbound traffic from the IP that originally belonged to host *A*2, but now belongs to host *Z*1 to the rest of the data center; and *d*) permits traffic on port 443 to host *A*3. The original policy allowing only port 443 to the IP of host *A*2, and the new policy allowing all traffic to the IP address of host *Z*1 are now in conflict.

10.4.1.2 Case Study 2: VPN Services

In a multi-tenant hosted data center, the provider could have layer-3 rules in place to prevent certain tenants from sending traffic to one another for monetization, compliance or regulatory reasons or even due to technical reasons. Hosts in two different tenant environments, Tenant *A* and Tenant *B*, can establish a layer-2 tunnel (either as a host-to-host tunnel or a site-to-site tunnel) between themselves to do single hop communication or to encrypt communication between them as shown in Figure 10.4. If another application running on the controller inserts policies to implement DPI, all traffic originating from

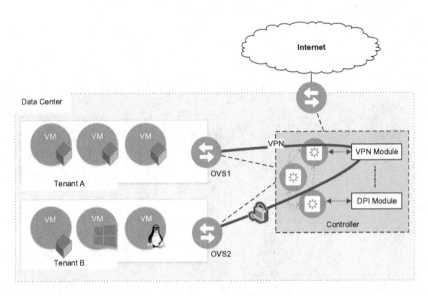

FIGURE 10.4
Policy conflicts caused by different applications in an SDN-based cloud.

Tenant *A* destined to Tenant *B* will be dropped, since they are encrypted and fail the DPI standards. Clearly, there is an inherent conflict between flow rules inserted by different applications running on the SDN controller, leading to a shoddy user experience.

10.4.1.3 Case Study 3: Load Balancing and IDS

As introduced in Section 10.4 and similar to the scenario in Case Study 2, consider an SDN-based data center environment where a load balancing application as well as an IDS application run on the SDN controller. Upon detecting intrusions, the IDS could implement a countermeasure that offloads traffic from the compromised node. However, the load balancing application which routes new connections based on their active load might start redirecting new traffic to the compromised node, since the system would infer that the compromised node has the least amount of load.

10.5 Flow Rule Conflicts

10.5.1 Problem Setup

When a packet arrives at an OVS, its match fields are compared to the match fields of the rules in the flow table. There are multiple ways a rule could be selected, namely: *a*) First match, where the first rule that matches the specified fields of the packet is selected; *b*) Best match, where the entire firewall rule set is examined to determine the rule that provides the tightest bounds to the specified fields; *c*) Deny take precedence, where any rule with a deny action is automatically preferred over other actions; and *d*) Most/Least specific take precedence, where the rule with the most/least specific match for the match fields [39]. The first match selection is by far the most prevalent way to select a matching flow rule. Here, we assume all selections to be based on first match selection, with rules ordered by priority. When multiple rules with the same priority exist, the newest rule has precedence.

10.5.2 Conflict Classes

Consider a flow table F containing rule set $\{r_1, r_2, \ldots, r_n\}$. We can represent a flow rule r_i using the tuple $(p_i, \epsilon_i, \zeta_i, \eta_i, \rho_i, a_i)$, where *a*) p_i is the priority. *b*) ϵ_i is the frame space of the rule. *c*) ζ_i is the packet space of the rule. *d*) η_i is the segment space of the rule. *e*) ρ_i is the OSI layer-4 protocol. *f*) a_i is the action set for the rule.

For all devices, including SDN devices or traditional firewalls, we deal with two main problems:

- Packet Classification Problem: In a firewall with rule set R, for an incoming packet Π_{in} with address 6-tuple n_{in} and protocol ρ_{in}, the packet classification problem [86] seeks to find out the set $R_m \subseteq R$ where $R_m = \{r_i \mid (r_i \in R) \wedge (n_i = n_{in}) \wedge (\rho_i = \rho_{in})\}$. The problem can be further extended to determine rule $r_x = (p_x, n_x, \rho_x, a_x) \in R_m$ such that $p_x > p_y \; \forall \; r_y \in R_m$.

- Conflict Detection Problem: The conflict detection problem [86] seeks to find rules r_i, r_j such that $r_i, r_j \in R$ and $(n_i = n_j) \wedge (\rho_i = \rho_j) \wedge (a_i \neq a_j \vee p_i \neq p_j)$.

We formally define the set operations on addresses at each OSI layer. Let $\xi \in \{\epsilon, \zeta, \eta\}$ be a 2-tuple (ξ_s, ξ_d) denoting an address at OSI layer-2, layer-3 or layer-4, with subscript s denoting the source address and d denoting the destination address. Then the following definitions apply.

Definition 7 $\xi_i \subseteq \xi_j$ *if and only if they refer to the same OSI layer, and* $\xi_{si} \subseteq \xi_{sj} \wedge \xi_{di} \subseteq \xi_{dj}$.

Definition 8 $\xi_i \nsubseteq \xi_j$ *if and only if they refer to the same OSI layer, and* $\xi_{si} \nsubseteq \xi_{sj} \vee \xi_{di} \nsubseteq \xi_{dj}$.

Definition 9 $\xi_i \subset \xi_j$ *if and only if they refer to the same OSI layer, and* $(\xi_{si} \subset \xi_{sj} \wedge \xi_{di} \subseteq \xi_{dj}) \vee (\xi_{si} \subseteq \xi_{sj} \wedge \xi_{di} \subset \xi_{dj})$.

Definition 10 *Address Intersection* $\xi_i \cap \xi_j$ *produces a tuple* $(\xi_{si} \cap \xi_{sj}, \xi_{di} \cap \xi_{dj})$ *if and only if* ξ_i *and* ξ_j *refer to the same OSI layer.*

Definition 11 *Conflict detection problem [86] seeks to find rules* r_i, r_j *such that* r_i, $r_j \in R$ *and* $(n_i \cap n_j \neq \emptyset) \wedge (\rho_i = \rho_j) \wedge (a_i \neq a_j \vee p_i \neq p_j)$.

Definition 12 *Flow rule address space* $n_i \subseteq n_j$ *iff* $\epsilon_i \subseteq \epsilon_j \wedge \zeta_i \subseteq \zeta_j \wedge \eta_i \subseteq \eta_j$.

Since flow rules in an SDN-based cloud environment are clearly a super-set of rules in a traditional firewall environment, work on flow rule conflicts are an extension of the work on firewall rule conflicts. While several works have classified firewall rule conflicts [86,108,175,293]; the seminal work by Al-Shaer and Hamed [23] is often used to classify firewall rule conflicts in a single firewall environment. The classifications used in the work of Al-Shaer and Hamed [23] are extended to formally classify flow rule conflicts, and further adapted to suit a distributed environment.

Knowing that OpenFlow specifications clarify that if a packet matches two flow rules, only the flow rule with the highest priority is invoked, the classification of different conflicts in SDN environments are detailed in the remainder of this section. The conflict classification is visually represented in Figure 10.5 and Figure 10.6. Figure 10.5 shows the address space overlap and flow rule conflicts for rules with different priorities, and Figure 10.6 shows the address space overlap for flow rules with the same priority.

10.5.2.1 Redundancy

A rule r_i is redundant to rule r_j iff: *a*) address space $n_i \subseteq n_j$; *b*) protocol $\rho_i = \rho_j$; and *c*) action $a_i = a_j$. For example, consider rules 1 and 2 from Table 10.2, shown below for easy reference. Rule 2 has an address space that is a subset to the address space of rule 1, with matching protocol and actions. Hence, rule 2 is redundant to rule 1. Redundancy does not pose a serious issue, but instead, is more of an optimization and efficiency problem.

```
<flow_id = 1> priority = 51, nw_src = 10.5.5.0/24, nw_dst = 10.1.1.63,
nw_proto = 6, actions = output : 3

<flow_id = 2> priority = 50, nw_src = 10.5.5.5, nw_dst = 10.1.1.63,
nw_proto = 6, tp_dst = 80, actions = output :3
```

10.5.2.2 Shadowing

A rule r_i is shadowed by rule r_j iff: *a*) priority $p_i < p_j$; *b*) address space $n_i \subseteq n_j$; *c*) protocol $\rho_i = \rho_j$; and *d*) action $a_i \neq a_j$. In such a situation, rule r_i is never invoked since incoming packets always get processed using rule r_j, given its higher priority. Shadowing is a serious issue since it shows a conflict in a security policy implementation [23]. For example, rule 4 has the same address space as rule 1, with the same protocol, but conflicting actions. But, the priority of rule 4 is higher than that of rule 1, which results in rule 1 never being invoked. Hence, rule 1 is shadowed by rule 4.

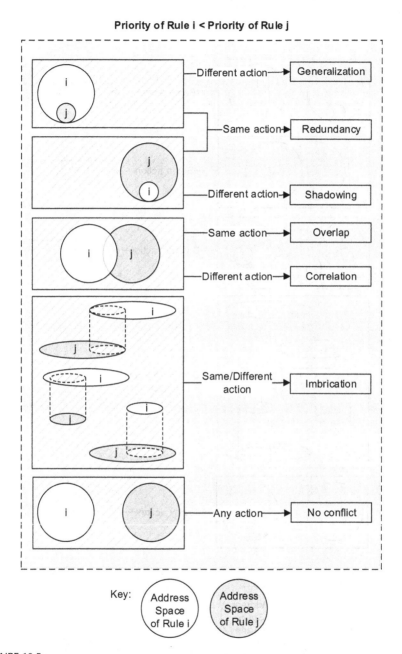

FIGURE 10.5
Address space overlap and flow rule conflicts for rules with different priorities.

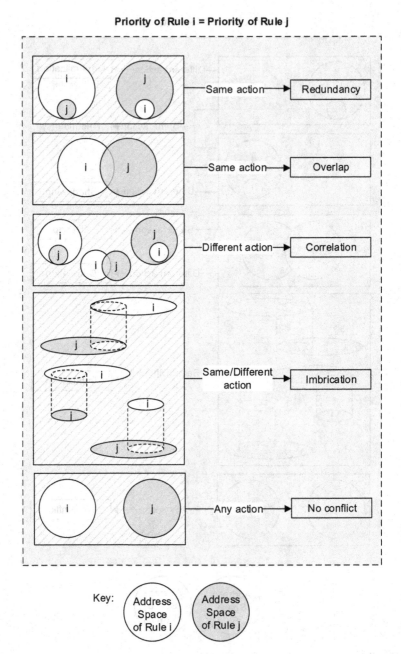

FIGURE 10.6
Address space overlap and flow rule conflicts for rules with same priority.

```
<flow_id = 1> priority = 51, nw_src = 10.5.5.0/24, nw_dst = 10.1.1.63,
nw_proto = 6, actions = output : 3

<flow_id = 4> priority = 53, nw_src = 10.5.5.0/24, nw_dst = 10.1.1.63,
nw_proto = 6, actions = drop
```

10.5.2.3 Generalization

A rule r_i is a generalization of rule r_j iff: a) priority $p_i < p_j$; b) address space $n_i \supseteq n_j$; and c) action $a_i \neq a_j$. In this case, the entire address space of rule r_j is matched by rule r_i [23]. As shown below, rule 1 is a generalization of rule 5, since the address space of rule 5 is a subset of the address space of rule 1, with the same protocols, but different actions. Note that if the priorities of the rules are swapped, it will result in a shadowing conflict. In traditional firewall management practices, it was common practice to add such rules for administrators to isolate a smaller portion of the traffic managed separately from a larger set of traffic.

```
<flow_id = 1> priority = 51, nw_src = 10.5.5.0/24, nw_dst = 10.1.1.63,
nw_proto = 6, actions = output : 3

<flow_id = 5> priority = 54, nw_src = 10.5.5.0, nw_dst = 10.1.1.63,
nw_proto = 6, actions = drop
```

10.5.2.4 Correlation

Classically, a rule r_i is correlated to rule r_j iff: a) address space $n_i \not\subseteq n_j \land n_i \not\supseteq n_j \land n_i \cap n_j \neq \emptyset$; b) protocol $\rho_i = \rho_j$; and c) action $a_i \neq a_j$ [23]. As shown below, rule 3 is correlated to rule 4.

```
<flow_id = 3> priority = 52, nw_src = 10.5.5.5, nw_dst = 10.1.1.0/24,
nw_proto = 6, actions = output : 2

<flow_id = 4> priority = 53, nw_src = 10.5.5.0/24, nw_dst = 10.1.1.63,
nw_proto = 6, actions = drop
```

Since multiple SDN flow rules can have the same priority, we make the following addition to the correlation conflict to satisfy requirements in an SDN environment: A rule r_i is correlated to rule r_j iff: a) priority $p_i = p_j$; b) address space $n_i \cap n_j \neq \emptyset$; c) protocol $\rho_i = \rho_j$; and d) action $a_i \neq a_j$. Thus, the correlation conflict now encompasses all policies that have the different actions, overlapping address spaces and the same priority. Scenarios where address spaces of two flow rules are subsets or supersets, which would have been categorized as generalization and shadowing in a traditional environment are classified as a correlation if the priorities of the two flows are the same. For example, in Table 10.2, rule 6 is correlated to rule 1.

```
<flow_id =1> priority =51 , nw_src =10.5.5.0/24 , nw_dst =10.1.1.63,
nw_proto =6, actions = output :3

<flow_id =6> priority =51 , nw_src =10.5.5.0/16 , nw_dst =10.1.1.63,
nw_proto =6, actions = drop
```

10.5.2.5 Overlap

A rule r_i overlaps rule r_j iff: *a*) address space $n_i \not\subseteq n_j \wedge n_i \not\supseteq n_j \wedge n_i \cap n_j \neq \emptyset$; *b*) protocol $\rho_i = \rho_j$; and *c*) action $a_i = a_j$. An overlap rule is similar to a correlation; but with the same action set. Note that the overlap conflict holds irrespective of the priority of the rules in question. This overlap can be seen between rule 6 and rule 7 in Table 10.2, shown below.

```
<flow_id =6> priority = 51, nw_src = 10.5.5.0/16, nw_dst = 10.1.1.63,
nw_proto = 6, actions = drop

<flow_id = 7> priority = 55, nw_src = 10.5.5.5, nw_dst = 10.1.1.0/24,
nw_proto = 6, tcp_dst = 0x03e8/0xfff8, actions = drop
```

10.5.2.6 Imbrication

The criteria discussed above does not cover all potential conflicts in SDN environments. Consider the case of flow rules where: *a*) only layer-3 header fields are used as a condition (rule 1-7 in Table 10.2); *b*) only layer-2 header fields are used as a condition for decision (rule 8); and *c*) only layer-4 header fields are used as a condition (rule 9). Even though using our definitions there is no overlap in address space, and hence there should be no conflict, a packet could match more than one of these rules. We classify such policy conflicts as imbricates, and address them by introducing the concept of *reconciliation* which maps all headers to the same layer. Currently, all cross-layer conflicts are classified as imbrication. They are examined in further detail below.

10.5.3 Cross-layer Policy Conflicts

As opposed to a traditional network, flow rules in SDN, could have matches on multiple header fields, thereby resulting in indirect dependencies. For example, consider traffic originating from Host A destined to Host B in Figure 10.7. This flow would clearly match both the flow rules shown in Listing 10.1. Rule with cookie value 0x2b0b would match on the layer-2 source, and layer-3 destination address; while rule with cookie value 0x2b3a would match on the layer-2 source, and layer-2 destination address. Since both rules have the

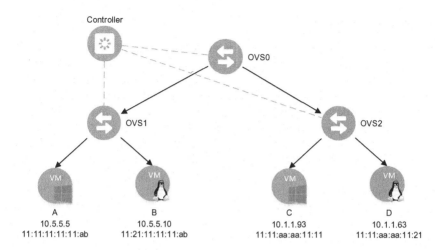

FIGURE 10.7
Cross-layer flow rule conflict in SDN environments.

same priority, the action taken by the controller would be inconsistent. As mentioned earlier, since there is no direction in the specification on how to deal with such a scenario, different controllers may deal with these conflicts in a different manner. A flawed approach to tackle this problem would be to expand the address space from layer-3 and layer-4 to include layer-2 addresses, and determine rule conflicts as in a traditional environment. However, since there exists an indirect dependency between the layer-2 and layer-3 addresses, an apples-to-apples comparison is impossible. Moreover, flow rules could exist that do not specify all the header fields adding another wrinkle.

Further, such conflicts between rules based on addresses over multiple OSI layers are more complex than the other conflict classifications, since they are transient in nature. For example, the mapping between a layer-2 MAC address and layer-3 IP addresses in Figure 10.7 might result in a conflict between two flow rules at time t_1 in the layer-3 address space. But if the IP-MAC address mapping changes, there may not be an address space overlap between the two rules at time t_2. This makes imbrication conflicts hard to find and even harder to resolve.

LISTING 10.1 FLOW RULE CONFLICT BASED ON ADDRESSES IN DIFFERENT OSI LAYERS

```
cookie = 0x2b0b,  duration = 926.421s,  table = 0,  n_packets = 1378,
n_bytes = 271308,  idle_age = 77,  priority = 100  dl_type = 0x800
dl_src = 11:11:11:11:11:ab  nw_dst = 10.211.1.63  actions = NORMAL

cookie = 0x2b3a,  duration = 949.733s,  table = 0,  n_packets = 622,
n_bytes = 957,  idle_age = 144,  priority = 100,  dl_type = 0x800
dl_src = 11:11:11:11:11:ab  dl_dst = 11:11:aa:aa:11:21  actions = drop
```

10.5.4 Traffic Engineering Flow Rules

Traffic engineering (TE) generally includes analysis of network traffic to enhance performance at operational and resource levels [32]. In data center and cloud service provider environments, QoS and resilience schemes are also considered as major TE functions, especially since several applications not only have bandwidth requirements, but also require other QoS guarantees [19]. Given the holistic network view that the SDN controller possesses, TE mechanisms in SDN can be much more efficient and intelligent, when compared to traditional IP-based mechanisms. Research on SDN based TE has tackled the tradeoffs between latency and load balancing, focusing on: *a*) controller load balancing[2] [87,122,159,267,289]; *b*) switch load balancing [20,43,70,71,291]; and *c*) the use of multiple flow tables. Our focus is determining how any implementation of TE functions in SDN environments using flow rules might conflict with security policies. We steer clear of considering controller and switch load balancing issues dealing with TE, and look at how implementing TE policies that direct traffic along certain paths, and implementing QoS for certain flows might interfere with security policy concerning the same flows.

OpenFlow specifications enable packets belonging to certain flows to be directed to a queue of an egress port. However, using queues to implement QoS requires that some configuration be done on the switches themselves, in addition to the controller. The snippet below shows creation of a QoS queue on an OVS that rate limits the maximum rate of this QoS policy to 1 Mbps, while setting the maximum rate to 5 Mbps.

```
$ ovs - vsctl set port eth0 qos = @newqos -- --id = @newqos create
  qos type = linux - htb other - config :max - rate = 1000000
  other - config :max - rate = 5000000
```

Further QoS-related additions in OpenFlow enable the setting of rate limiting functions. These utilize meter table entries, which define per-flow meters that measure rate of packets assigned, and enable controlling that rate. Meters are associated directly with flow entries, as opposed to different queues of egress ports, and contain: *a*) an identifier; *b*) the specified way to process the packet; and *c*) counters. Use of the meter table, as shown in Figure 10.8, to establish a game theory-based security framework was demonstrated by Chowdhary et al. [63], wherein the `rate` sub-field of `band` field in meter table was used to establish rate limiting for non-cooperating actors. Their work shed light on novel ways to use TE to implement security.

Our approach to tackling conflicts while using either of these two QoS scenarios is abecedarian, wherein only the forward/deny aspect of the rule is considered in the detection and resolution of conflicts.

---

[2] Multiple controller scenarios are discussed in Chapter 3.

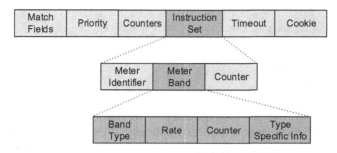

FIGURE 10.8
Meter band in OpenFlow specification.

10.6 Controller Decentralization Considerations

Choosing a decentralized control architecture is not trivial. There are several controller placement solutions, and factors such as the number of controllers, their location, and topology impact network performance [137]. Three major issues need to be elucidated [245] while determining the decentralization architecture:

- Efficient east- and westbound APIs need to be developed for communication between SDN controllers.
- The latency increase introduced due to network information exchange between the controllers needs to be kept to a minimum.
- The size and operation of the controller back-end database needs to be evaluated.

Since the key piece of information required for accurate flow rule conflict detection (and resolution, as will be described in Chapter 10) is the priority value of the flow rule p, the key challenge in extending flow rule conflict resolution from a single controller to a distributed SDN-based cloud environment lies in associating *global priority* values to flow rules. Definition 13 defines global priority.

Definition 13 *A global priority p' of a rule r_i is value in the range [1, 65535], as determined by weighing the priority value p by the rule origination point's position in the global distribution scheme. Alternately, p' could be obtained using a static mapping scheme from p.*

To illustrate Definition 13, consider flow rule as shown with a priority value of 51. If this flow rule originated in a controller or application with a weight of 2, its global priority would be 102.

```
priority = 51, nw_src = 10.5.50.5, nw_dst = 10.1.1.1,
actions = output : 3

global_priority = 102, priority = 51, nw_src = 10.5.50.5,
nw_dst = 10.1.1.1, actions = output : 3
```

The strategies to associate these global priority numbers to flow rules in different decentralization scenarios differ drastically. We classify five different multiple controller scenarios, and the global priority assignment logic followed by our framework for each of them.

10.6.1 Clustered Controllers

The clustered SDN controller is the simplest of the multiple controller environments. It is ideal for smaller networks, where one controller can process all events and run all applications. Clustering adds a layer of defense against the controller being a single point of failure by having one or more controllers in an active/standby scenario. Since all the controllers run the same applications and communicate with all the data plane devices, the global priority assigned to the rules would be equal to the priority of the rule.

Flow rule conflict classification and resolution in clustered controllers is handled exactly like in a single controller environment, owing to a lack of partitioning of data plane devices or applications. Hence, discussion of clustered controllers as a decentralization strategy is limited in the rest of this chapter.

10.6.2 Host-based Partitioning

Host-based partitioning is most like a traditional layered network architecture, where an SDN controller handles the functionalities of an access layer switch, combined with the intelligence of a router and access control server. The SDN-based cloud environment is separated into domains, where each domain is controlled by a single controller. As shown in Figure 10.9, the tenant infrastructure in multi-tenant data center environment could be considered a domain that is handled by one controller. All the controllers present in the environment would maintain global knowledge of the environment by communicating with each other using east-west communication APIs.

Running on the assumption that the controller knows best about the main it is responsible for, flow rules which contain match conditions with addresses *local* to the controller are preferred. For example, the rule with cookie value 0xa added onto Controller 1 permits DNS traffic into host 10.1.1.5, which we assume is an address assigned to Tenant *A*. If the rule with cookie value 0xb is added on Controller 2, the two conflicting flow rules will be known to all the controllers owing to the controllers sharing their information.

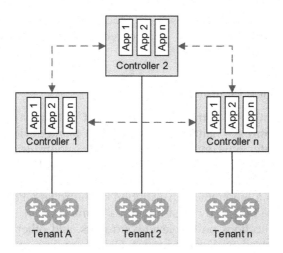

FIGURE 10.9
Host-based partitioning.

```
cookie = 0xa, priority = 100, nw_dst = 10.1.1.5, nw_proto = 17,
udp_dst = 53, actions = output : 1
cookie = 0xb, priority = 100, nw_dst = 10.0.0.0/8, nw_proto = 17,
udp_dst = 53, actions = drop
```

To help select the rule most applicable to the tenant, we assign weights to the flow rules such that the ones originating from the controller assigned to the specific domain (Controller 1, in our example) is considerably higher. The weight itself is dependent on the environment, and can be assigned by an administrator. Assuming a weight of 10 for the local controller, we now have global priorities as shown in the modified flow rules below. The global priority value can then be used for conflict resolution.

```
global_priority = 1000, cookie = 0xa, priority = 100, nw_dst = 10.1.1.5,
nw_proto =17, udp_dst = 53, actions = output : 1
global_priority = 100, cookie = 0xb, priority = 100, nw_dst = 10.0.0.0/8,
nw_proto = 17, udp_dst = 53, actions = drop
```

Host-based partitioning is popular in several cloud deployments, owing to its simplicity. In DragonFlow [4], for example, an instance of the controller runs on every compute node in the OpenStack cloud. Partitioning in DragonFlow is based on the node it runs on, and is not purely tenant-based. Thus, an instance of DragonFlow would manage the OVS and flow rules that are associated with hosts running on the same compute node. The different DragonFlow instances in the cloud share information by communicating with a shared back-end database. Partitioning schemes such as those employed by Dragon-Flow are intuitive and most like decentralization strategies used in traditional environments.

10.6.3 Hierarchical Controllers

Hierarchical controller distribution is a variant of host-based partitioning, where some controllers handle a subset of data plane devices, while others only communicate with control plane devices. The controllers that communicate with the data plane devices can be thought of as leaf-level controllers, while higher-level controllers communicate solely with other controllers. Figure 10.10 shows a hierarchical distribution of controllers. Further, the partitioning may not be strictly host-based, as administrators could decide to run certain applications on leaf-level controllers, and other applications on higher-level controllers. For example, a DHCP application could reside on the leaf controller while a NAT application could reside on the root controller.

Since higher-level controllers do not communicate with data plane devices, except in cases when leaf-level controllers fail, control channel communication is streamlined. Leaf-level nodes can obtain global information by communicating with the higher-level controller, eliminating the need to talk with every other leaf controller. Since the root controller would have holistic knowledge of the environment, in case of conflicts flow rules originating from the root controller are preferred.

Revisiting the example from host-based partitioning scheme, consider that the rule with cookie value 0xa added onto Controller 1, while the rule with cookie value 0xb is added on Controller 0. While Controller 1 might still

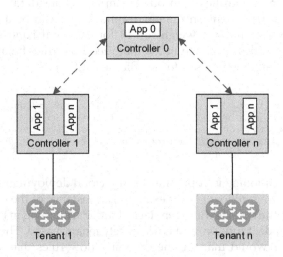

FIGURE 10.10
Hierarchical controller distribution.

permit DNS traffic into the host at IP 10.1.1.5, the root level controller might have an IDS application running on it that detected a DDoS attack, and dropped all DNS traffic to the devices on 10.0.0.0/8 subnet. Once again, we use a weight of 10, but this time for the rule on Controller 0, thereby ensuring the conflict resolution algorithm drops the attack traffic.

```
global_priority = 100, cookie = 0xa, priority = 100, nw_dst = 10.1.1.5,
nw_proto = 17, udp_dst = 53, actions = output : 1
global_priority = 1000, cookie = 0xb, priority = 100, nw_dst = 10.0.0.0/8,
nw_proto = 17, udp_dst = 53, actions = drop
```

Hierarchical controllers may also be used if there is a specific structure to the network, such as a two-tier structure. In such situations, hierarchy at the controller level can help to manage flows for different tiers in the network. However, coordination needs to be addressed to ensure efficiency in flow management.

10.6.4 Application-based Partitioning

Application-based partitioning, as shown in Figure 10.11 implements decentralization by having different applications run on different instances of the controller. As with host-based partitioning, the flow rules generated by the different applications would be known to the other controllers using the east-west communication APIs. Each data plane device in this scenario would communicate with every controller in the environment.

Associating global priority values in application-based decentralization is straightforward. It could be done by assigning a weight to each application [218], and is also used to generate the global priorities of flow rules. For

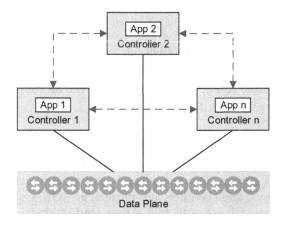

FIGURE 10.11
Application-based controller decentralization.

example, consider Controller 1 with security applications running on it, and Controller 2 with QoS and traffic shaping applications running on it. If security applications are prioritized with a higher weight than traffic shaping applications, flow rules with the same priority generated by applications on Controller 1 and Controller 2 will end up with the rule generated by Controller 1 having a higher global priority.

An alternate strategy to assign global priority values would be to allocate ranges for flow rules created by applications. For example, it could be decided that any NAT rule generated by the NAT application on the controller must be within a priority of 40,000 and 42,000. Thus a global priority for a NAT rule would be generated by mapping the priority originally in the range [1, 65535] to a global priority in the range [40000, 42000].

10.6.5 Heterogeneous Partitioning

In a heterogeneous decentralized environment, appealing aspects of each of the above decentralization scenarios are combined to obtain the optimal situation for meeting the requirements. Careful consideration needs to be taken to identify the priorities of applications and controllers before deployment, to have a conflict resolution strategy.

10.7 Flow Rule Conflict Resolution

In this section, the considerations for flow rule conflict resolution are discussed. First, the flow rule conflict types discussed earlier are categorized based on their difficulty of resolution. A conflict resolution model that addresses automatic resolution of the flow rule conflicts is presented next.

10.7.1 Conflict Severity Classification

Based on their potential for causing damage, as well as their difficulty to resolve, the flow rule conflicts formalized above can be classified into the following tiers:

10.7.1.1 Tier-1 Conflicts

Imbrication conflicts stem from flow rules using addresses in multiple layers. Since mapping between different layers is of transient nature in dynamic SDN environments, these conflicts are transient as well. Any resolution technique used to resolve these conflicts is, at best, made taking the current system state into account. Such a resolution might induce the system to be in a highly compromisable state at a future time, which could be exploited by attackers.

To illustrate, consider once again the topology in Figure 10.7. If a layer-2 policy and a layer-3 policy that constrain traffic between hosts A and D are present, one of the conflict resolution strategies might select the layer-2 flow rule. However, if the mapping between the layer-2 and layer-3 addresses changes, the conflict resolution decision might be rendered invalid.

10.7.1.2 Tier-2 Conflicts

Conflicts classified as *generalization* and *correlation* stem from overlapping address spaces and incompatible actions. These conflicts stem from attempts at combining and oversimplifying flow rules. By making the address spaces used in the flow rules as fine-grained as possible, tier-2 conflicts can be eliminated.

10.7.1.3 Tier-3 Conflicts

Conflicts categorized as *redundancy* and *overlap* result from rules with overlapping address spaces but the same resulting action. The *shadowing* conflict stems from rules which are never invoked. In case of redundancy and overlap, the action remains the same, so choosing any either flow rule would result in the same action on the packet. Shadowed rules are never invoked owing to their lower priority. Thus, we contend that while it is not ideal that these conflicts exist in the system, their presence in the system is not a security threat but an optimization issue.

10.8 Conflict Resolution Model

The different flow rule conflicts can be broadly categorized into **Intelligible** and **Interpretative** conflicts. The resolution strategies for each of these two categories are markedly different, and are detailed in the remainder of this section. Tier-1 and Tier-2 conflicts are interpretative in nature, while Tier-3 conflicts are intelligible in nature.

10.8.1 Intelligible Conflicts

Flow rules that conflict with each other in the Redundancy and Overlap classifications all have the same action: they can be resolved without the loss of any information. Rules that have shadowing conflicts can simply be removed, without affecting any packet. In other words, the resolution algorithm can guarantee that any packet that is permitted by the controller prior to resolving the conflict will continue to be permitted after conflict resolution. And similarly, any packet that is being blocked prior to conflict resolution will continue to be blocked after the conflict resolution is put in place. Intelligible conflicts are resolved easily by eliminating the rules that are not applied, or by combining and optimizing the address spaces in the rules to avoid the conflict [150].

It could be argued that creative design of rules by administrators result in flow rules that deliberately conflict to optimize the number of rules in the flow table, especially when it comes to traffic shaping policies. However, such optimization strategies stem out of legacy network management techniques, and do not hold true in dynamic, large-scale cloud environments where the flow table enforcing the policies in the environment could have millions of rules.

10.8.2 Interpretative Conflicts

Conflicts that fall into Generalization, Correlation and Imbrication classification cannot be intuitively resolved without any loss of information, and are interpretative in nature. As opposed to intelligible conflicts, it is not guaranteed that any packet permitted by the controller prior to resolving the conflict will be permitted after conflict resolution. Since interpretative conflict resolution is lossy in nature, the resolution strategies are *not* a one-size-fits-all and need to be adapted per the cloud environment in question. Removing these conflicts is a complex problem [24].

A few different resolution strategies that could be applied to resolving these conflicts are discussed below. The global priority of the rule is assigned depending on the controller decentralization strategy discussed in Chapter 3. Resolution strategies for Tier-1 conflicts are shaky at best. Since these conflicts are transient in nature, an additional decision needs to be made as to the time duration for which the conflict resolution strategy is valid. We look at four different strategies.

10.8.2.1 Least Privilege

In case of any conflict, flow rules that have a deny action are prioritized over a QoS or a forward action. If conflicts exist between a higher and lower bandwidth QoS policy, the *lower* QoS policy is enforced. The least privilege strategy is traditionally the most popular strategy in conflict resolution [239].

10.8.2.2 Module Security Precedence

Since flow rules in an SDN-based cloud environment can be generated by any number of modules that run on the controller, an effective strategy that can be put in place is to have a security precedence for the origin of the flow rule [218]. Thus, a flow rule originating from a security module is prioritized over flow rule from an application or optimization module. The weighted global priorities are calculated as discussed in the application-based partitioning scheme discussed earlier. Table 10.3 shows sample precedence and associated global priority weight values for a few generic applications that might run in an SDN-based cloud.

TABLE 10.3

Security Precedence Priority Multiplier Example

Application	Precedence	Global Priority Weight
Virtual Private Network	1	3
Deep Packet Inspection	2	2.5
Network Address Translation	3	2
Quality of Service	4	1.5
Domain Name Service	5	1

10.8.2.3 Environment Calibrated

This strategy incorporates learning strategies in the environment to make an educated decision on which conflicting flow rule really needs to be prioritized. Over time, if a picture can be formed about the type of data that a certain tenant in a multi-tenant data center usually creates/retrieves, or of the applications and vulnerabilities that exist in the tenant environment, or of the reliability of the software modules inserting the flow rule; the conflict resolution module may be able to prioritize certain flow rules over others. However, these techniques falter while dealing with a dynamic cloud.

A more deliberate approach might involve quantitative security analysis of the environment with each of the conflicting rules, and picking the safest option. Metrics originally proposed by Joh and Malaiya [138] and validated by Lippmann et al. [170] provide a quantitative measurement of the probability of the Cyber Key Terrain (CKT) [226] being compromised. Interpretative conflict resolution could be as simple as determining which of the conflicting policies would reduce the compromise probability.

10.8.2.4 Administrator Assistance

Administrators that are willing to give up automatic conflict resolution have the option to resolve conflicts manually, so they can judge each conflict independently. Visual assistance tools incorporated as part of the Brew framework assist the administrator make a decision and are detailed in [201].

Summary

This chapter described the problem of policy conflicts, and their prevalence especially in distributed environments. How security policies map to Open-Flow rule conflicts is studied. The challenges that are brought about by a distributed environment are discussed. All potential flow rule conflicts are categorized, and techniques to resolve said conflicts are examined. That in of itself enables us to obtain a solid foundation to having a platform that can ensure a conflict-free security policy implementation across a distributed SDN environment.

For continuing work based on the foundation laid out in the chapter, integrating results from studies that incorporate stateful functionality into the SDN environment would be an interesting avenue to explore. Evolving from a pure packet filter-based security application to one that can have rules based on connection state would greatly enhance the effectiveness of security policy that can be implemented. Research from the traditional networking environment suggests this may not be very complex [53].

Verifying that new flow policies adhere to organizational security policies is a twofold problem: *a*) establish that the newly generated flow rules do not conflict with existing flow rules; and *b*) ensure that the conflict-free flow rule table adhere to the high-level organizational policies. To ensure compliance with higher level organizational policies, the work described here needs to be adapted/extended to work in the area of regulatory compliance. Such work usually uses a policy specification language based on a restricted subset of First Order Temporal Logic (FOTL) which can capture the high-level requirements and encode what adherence to the policies mean [65]. Combining this work with FOTL would greatly increase its future applications.

Further, considering flow rule optimization based on rule positioning and examining adaptive prioritization of rules would be interesting. Including role-based and attribute-based policy conflicts is a natural extension of this work. And finally, using machine learning algorithms to identify MTD timer thresholds and resolve interpretative conflicts could be a fruitful problem to solve.

11

Intelligent Software-Defined Security

Software-Defined Security is an approach to implementing, managing and controlling information security in a computing environment using the software. The security components such as intrusion detection, access control, network segmentation are automated and managed through software. There is very limited or no hardware-based security dependence. The software-defined networking framework helps in managing and orchestrating the security needs of an organization in an intelligent fashion.

The Intelligent Software-Defined Security (ISDS) that we discuss in this chapter comprises of key properties of an intelligent software system such as situation awareness, self-healing, end-to-end monitoring, network analytic capability, and feedback mechanism to dynamically reconfigure the network in case of any compromising activity.

We discuss some important architectural considerations in the application of Machine Learning (ML) and Artificial Intelligence (AI) in security in Section 11.1. Different ML and AI techniques such as neural networks, expert systems, learning mechanisms along with their security applications have been briefly discussed in this section. Additionally, we use an intrusion detection system (IDS) as an example to showcase the application of intelligence in the field of security in Section 11.1. SDN-based intelligent security design that incorporates ML and AI have been described in Section 11.1.4. Section 11.2 has been dedicated to the study of advanced persistent threats (APTs). The difference between traditional attacks and APTs, examples of most notable APT events and vulnerabilities have been discussed in this section. The techniques used in detection and mitigation of APT have been discussed in Subsection 11.2.4. Subsection 11.2.5 describes the use of SDN based microsegmentation and defense-in-depth security to disrupt the propagation of APTs. Section 11.3 has been dedicated to the study of problems associated with the application of intelligence in security such as variance in network traffic, high cost or errors because of incorrect attack prediction.

11.1 Intelligence in Network Security

11.1.1 Application of Machine Learning and AI in Security

Machine Learning (ML) is a field of computer science that utilizes statistical techniques to help computers learn using the available data, without being explicitly programmed [237]. Artificial Intelligence (AI) [217] on the other hand is defined as a study of intelligent agents. The intelligent agents perceive their environment and take necessary actions that maximize the chances of achieving the desired goal for the agent. AI is applied when a machine mimics cognitive functions similar to the human brain such as *learning* and *problem solving*. Commercial applications of AI have been pioneered by technology giants such as Google (search engine) and Facebook (news feed) [167].

The user behavior data is collected, cleaned and analyzed by the organizations using big data analytic frameworks, ML and AI algorithms, to predict the usage trends and derive more business value from their data. AI and ML help in bolstering the cybersecurity infrastructure by using complex and sophisticated applications to detect stealthy attack patterns. The adaptive security framework built using AI and ML can automatically detect, analyze and prevent attacks. Some proactive defense mechanisms offered by AI include data deception techniques which can trick the attackers into interacting with Honeypots and Honeynets [257] instead of legitimate services. Although AI can itself introduce new threat vectors when it is dependent upon interfaces within and across the organization, and attackers can utilize the AI techniques to learn and target the ISDS, we consider such use cases beyond the scope of this chapter.

11.1.2 Intelligent Cybersecurity Methods and Architectures

Artificial Intelligence, Machine Learning, and Data Mining have several methods and architectures that find practical use in different fields of cybersecurity as discussed by Tyungu et al. [270]. We discuss some of these methods in this subsection.

11.1.2.1 Neural Networks

Neural Networks (NN) mimic the human cognition capabilities using a network of artificial neurons interacting with each other. Simplest neural networks are known as Perceptron [231]. Some applications of NN include attack pattern recognition, intrusion detection, and prevention. With the support of suitable hardware and graphics processors, NNs can achieve high-speed detection rate, which can help in scaling security solution on a large network.

11.1.2.2 Expert Systems

Expert System handles the task of modeling human reasoning with the aim of finding solutions to questions in application domains such as finance, medical diagnosis or cybersecurity. Two important parts of expert systems include i) *Knowledge Base* for storing the expert knowledge about a specific domain, and ii) *Inference Engine* for deriving answers based on the knowledge base and additional information. *Expert system shell* consists of an empty knowledge base and inference engine. The knowledge base must be filled with the required knowledge before an expert system can utilize it. Additionally, expert systems can have simulation capabilities to simulate various instances of cyber attack and defense. *Security Planning*, which involves selection of suitable security measures and optimal usage of limited security resources is an application area of expert systems in cybersecurity.

11.1.2.3 Intelligent Agents

Intelligent Agents are software or hardware components that exhibit some properties of intelligent behavior such as proactiveness, situation awareness, and understanding communication language from other intelligent agents. Intelligent agents can cooperate and provide defense against cyber attacks such as distributed denial of service (DDoS) as discussed by Konteko et al. [161, 162]. Multi-agent systems can also be utilized in building hybrid and distributed intrusion detection systems [57, 111].

11.1.2.4 Learning

Learning is the task of improving the intelligent systems by rearranging the knowledge base or improving the inference engine. Machine Learning techniques can vary in complexity from simple learning task, which involves learning some parameters - *Parameter Learning* to more sophisticated learning concepts such as grammar, concepts and user behavior - *Symbolic Learning*. Another method of classifying the task of learning is *Supervised Learning* and *Unsupervised Learning*. When large-scale cybersecurity datasets are present such as DDoS logs, user activity, and system process data, unsupervised learning can be very helpful. Neural Networks and Self Organized Maps (SOMs) [34, 276] utilize unsupervised learning for intrusion detection and threat analysis.

11.1.2.5 Search

Search problem in AI involves selecting the best solution from a list of candidate solutions. Another aspect of search is the utilization of additional knowledge to guide the search and improve the efficiency of the search. Search algorithms such as min-max, Stochastic Search, and $\alpha\beta$-pruning can be utilized to solve security problems in an optimal fashion [205]. In a cybersecurity setting, the attack-defense scenarios can be considered a static or Dynamic Game between the attacker and defender. The search algorithm can help the defender select a security configuration that guarantees minimum and maximum returns for the current security setting.

11.1.2.6 Constraint Solving

Constraint Solving, or Constraint Satisfaction Problem (CSP), involves solving a set of constraints - equations, inequalities, and logical statements for a problem. Planning problems in AI can be represented as CSP. The search problems can sometimes be difficult to solve because of a large amount of search required over the available data. The constraints restrict the search to a narrow dataset by taking into account information about a particular class of problem. Constraint solving can be used in situation awareness and decision support by utilizing a logic programming approach [210].

11.1.3 Application of AI in IDS

The Artificial Intelligence (AI) deals with problems such as searching for an efficient solution, feature selection, selection of most relevant feature set, which can help in reduction of data-set, performance overhead, and storage requirements. The expert system utilizes training instances to acquire knowledge. The systems can also utilize training instances to acquire knowledge. There are two main types of techniques based on training instances used in AI, i.e.,

1. **Rule Based Induction** derives the rules that are able to explain the training instances better than mathematical or statistical techniques.
2. **Classifier System** utilizes a set of training data in order to classify the future examples. Examples of a classifier system include *Neural Network* [104] and *Decision Tree* [157].

We analyze the survey of AI techniques employed in the Intrusion Detection System (IDS) [93]. An important characteristic desired in IDS is real-time analytic capability. Analysis of a huge volume of network traffic in a cloud environment may take several hours, at the end of which the predicted result may not be very useful. Therefore data reduction techniques can help in speeding up the prediction of intrusion events in real time.

11.1.3.1 Data Reduction

The data not very useful for intrusion detection can be filtered out using the *Filtering* process. The assumption is that user activity will have some notable trends, which can easily be detected and account for correlation in the input data. The useful data can undergo a second step, i.e., *Feature Selection*, which can be employed to eliminate data not containing desired features for intrusion detection. The process of feature selection can help in obtaining the feature, most indicative of misuse activity or distinguish between different types of misuse activities. A third step in the process of data reduction is *Clustering*. This process can help in finding hidden patterns in data and storing the characteristics of the entire cluster instead of actual data. There are several kinds of clustering techniques such as *Hierarchical Clustering* (generalization based), *Statistical Clustering* (probability of example being in a cluster), *Distance Clustering* (distance to establish membership in a cluster), etc.

11.1.3.2 Behavior Classification

The IDS can identify only a certain fraction of intrusion events correctly. Often the normal users are incorrectly flagged as malicious by the IDS (False Positive) and vice versa (False Negative). The AI techniques such as *Statistical Anomaly Detection* can help in improving the IDS classification performance. The expert systems encode the known attacks and IDS policies as a fixed set of rules. The user behavior is matched against the rules to determine the attacks. Expert systems incorporate rule encoding, which can be utilized to make conclusions regarding the information gathered by IDS. The presence of domain expertise in expert systems provides optimal quality of rules. The encoded rules can be past intrusion events, system vulnerabilities, and security policies. The anomaly detection component compares the attacker's behavior against the normal expected user behavior. Three distinct phases of IDS are:

1. Local information extraction.
2. Evolving background information from local abstraction.
3. Establishment of anomaly background boundaries.

Some operations that are performed on the raw data before behavior classification include smoothing of the raw data to eliminate outliers, data weighting to assign a higher weight to historical data than current data, blending of the behavior variations to establish a tolerance level for network anomaly detection. Some anomalous patterns can include resource usage variations, login session variations, directory access patterns, etc.

11.1.4 SDN-based Intelligent Network Security Solutions

The centralized command and control design of SDN provides flexibility, simplicity, and elasticity. The SDN application plane can be extended to develop many intelligent applications such as smart IDS/IPS, traffic anomaly classification, content popularity prediction, link/bandwidth testing application, path discovery application, etc. We discuss some security applications and frameworks that utilize SDN. An example of SDN-based intelligent network security solution can be seen in Figure 11.1. The application plane in SDN can have several smart applications such as path discovery, web request classification, traffic anomaly classification module, which can be implemented using SDN controller, that acts as a middle layer between application and data plane.

11.1.4.1 Topology Protection

SDN controller can be subjected to network topology-based attacks. The fundamental building block of SDN is the discovery of network topology. With a poisoned topology, the visibility of upper layer services and apps may be

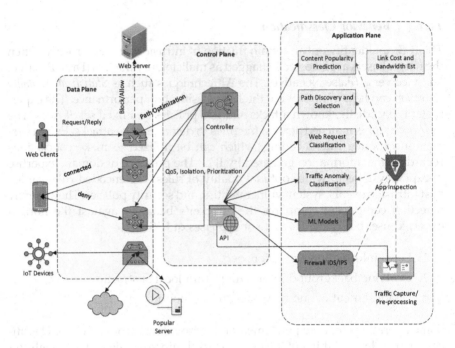

FIGURE 11.1
SDN-based intelligent network security solutions.

misled by the attacker, leading to hijacking, Man-in-the-Middle (MITM), and denial-of-service. The link discovery and host tracking service APIs in various SDN controllers such as NOX, POX, Floodlight, and Ryu can be subjected to these attacks. Some attack vectors that target network topology have been described below.

- **Host Location Hijacking:** OpenFlow controllers consist of a Host Tracking Service (HTS). The events such as host location migration can be identified by HTS by monitoring *Packet-In* messages. Once the host migrates to a new location, the HTS updates the host profile information. Existing OpenFlow controllers have very weak security for host location update service. OpenDayLight controller consists of API *isEntityAllowed*, which accepts every host location update. An adversary can tamper with host location information by impersonating the target host. The host location information is utilized by SDN controllers to make packet forwarding and routing decisions. Once the attacker is able to tamper with host location information, the traffic towards the host can be hijacked.

  As shown in Figure 11.2, the attacker can create fake packets with an identifier of the target host. Once the OpenFlow controller receives the

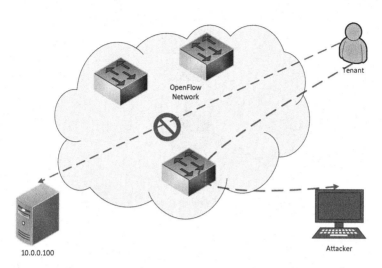

FIGURE 11.2
Attacker impersonating web server to hijack communication and phish users.

spoofed packet, the controller is tricked into believing that the host has migrated to a new location. In effect, any future traffic meant for the target host is hijacked by the attacker.

- **Link Fabrication Attack:** The OpenFlow controllers utilize Link Layer Discovery Protocol (LLDP) to discover the links between various switches. Additionally, OpenFlow Discovery Protocol (OFDP) and Link Discovery Service (LDS) are utilized by OpenFlow controller to construct network topology. The security flaws in the link discovery procedure can be exploited by a malicious attacker.

There are two security constraints defined as a part of LDS specification, i.e., 1) The integrity/origin on LLDP packets must be ensured in the link discovery procedure. 2) The propagation path of LLDP packets must only contain OpenFlow-enabled switches. The incorrect enforcement of these constraints by existing OpenFlow controllers opens up a window of opportunity for the attacker. The adversary can craft falsified LLDP packets or relay LLDP packets between switches in order to fabricate a fake internal link.

As shown in Figure 11.3, an adversary forges the internal link in a relay fashion. The adversary receives the LLDP packet from one target switch and repeats it to another switch without any modification. In effect, the adversary constructs a fake topology view to the OpenFlow controller, as if there is an internal link between two target switches. The dashed line in the figure is the actual traffic route, whereas the dotted line is the communication channel between the users in the view of the controller. The link fabrication attack can serve as a basis for further attacks such as DoS and MITM attack.

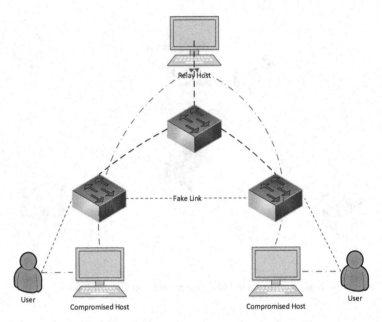

FIGURE 11.3
Link fabrication attack using LLDP relay.

- **Static Defense:** The static defense consists of configuring the host link and location information beforehand and manually verifying/ updating the information when required. The solution is however error-prone and difficult to scale on a large network.

- **Dynamic Defense against Host Location Hijacking:** The host location hijacking attack is successful in SDN environment because the OpenFlow controller fails to verify the host identifier when the host location is updated. The solutions that can be utilized to mitigate this problem include: 1) Authentication of host information using public-key infrastructure. When the host decides to change the location, the new location information can be embedded into unused packet field (VLAN or ToS), and packet encryption using the private key. This solution can, however, have scalability limitations on a large network. 2) The legitimacy of host migration can be verified by checking the pre-condition *Port_Down* before host migration and post-condition, i.e., checking that the migrated host entity is unreachable in the previous location after host migration is completed.

- **Dynamic Defense against Link Fabrication:** Link fabrication attacks can be mitigated by: 1) Adding additional authentication in the LLDP packet. The signature can be calculated over the semantics of the LLDP packet (DPID and Port number). 2) Switch property verification can be employed to check if the host resides inside LLDP propagation, e.g., traffic coming from each switch port can be inspected to check devices connected to the switch. Since LLDP can only transmit on the switch internal link ports and ports connected to the OpenFlow controller, this can help in preventing LLDP replay attacks.

11.1.4.2 SDN-based DoS Protection

The traditional ML-based models employed for protection against DoS attacks rely on network flow classification algorithms. The traffic logs are pre-processed to identify some statistics that can be utilized by ML models. This, however, introduces a lot of overhead in pre-processing and data collection. The SDN architecture consists of flow tables with a lot of packet header and counter details. The control plane can query the statistics related to the forwarded traffic anytime. The DoS detection algorithm consists of components such as *Data Collector*, *Feature Extractor* and *Flow Classifier*. Once the malicious traffic pattern has been detected, the bad traffic can either be dropped or forwarded to a Remote Triggered Black Hole Routing Component (RTBH).

Alshamrani et al. [26] use SDN framework to detect and prevent DDoS attacks. The authors use online and offline algorithms to train the ML model on 41 features defined on NSL KDD dataset [265]. Two different types of attacks, New Flow Attack and Misbehavior Attack, are detected using SMO-based ML algorithm. Additionally, the classification algorithm is able to classify traffic into DoS, probe, normal and privilege escalation attempt (R2L) traffic. The algorithm used, i.e., SMO achieves higher accuracy when compared to other classification algorithms such as J48 and Naive Bayes algorithm. Frameworks such as *Science DMZ* [62] also present a framework for detecting, analyzing and preventing network attacks using SDN.

11.2 Advanced Persistent Threats

Advanced Persistent Threats (APTs) are stealthy attacks mounted by a sophisticated group of attackers often sponsored by large organizations or governments to gain useful information about the target organizations. APT is a combination of three words [232] namely:

Advanced: The APT attacks are well funded and use *advanced* modes of operation and sophisticated tools as opposed to normal information discovery tools used by individual attackers. The advanced tools employ multiple attack vectors, and the target organization in case of APT is often a highly valued target.

Persistent: The group of attackers in case of APT are highly motivated and *persistent*. Once the attackers gain access into the system, they try to gain access to connected systems without raising security tool alarms. The attackers employ several evasive techniques and follow *"slow and low"* approach to increase the chances of success.

Threats: The *threat* in case of APT attack is a loss of data or critical information that can cause disruption in the normal operations of an organization, loss of reputation and mission-critical information. These threats are difficult to detect, and require sophisticated defense mechanisms to detect and prevent.

According to the National Institute of Standards and Technology (NIST) [154], an APT attacker: (i) pursues its objectives repeatedly over an extended period of time; (ii) adapts to defenders' efforts to resist it; and (iii) is determined to maintain the level of interaction needed to execute its objectives. And the objectives include exfiltration of information, undermining or impeding critical aspects of a mission or program through multiple attack vectors.

11.2.1 Traditional Attacks vs. APT

APT is often loosely defined currently, and used by organizations as an excuse when they fail to defend themselves against a targeted attack. On the other hand, attackers have in some cases used goals as part of APT attacks not well defined by NIST and other similar organizations. Thus, there is a need to reconsider the definition of APT, and include other attack vectors and attack goals as part of the APT definition. It is important to distinguish APT from traditional attacks by considering some of the questions like:

1. Can the attack be prevented in one or more way?
2. Does the attack require a great deal of adaptation by the attackers?
3. Does this attack exhibit novelty in its variants and is difficult to detect by traditional means?

If the answer to Q1 is *False* and answers to Q2,3 is *True*, the attack can be classified as APT. Table 11.1 considers some of the characteristics that can be used to distinguish normal attacks from APT.

11.2.2 APT Attack Model

APT attacks are well planned and highly organized in order to increase the probability of attack success. In order for the attack to be successful, APT is performed in multiple stages. Attack Trees described by Schneier et al. [240] are a useful tool to study Multi-Stage Attacks. APTs can be modeled using

TABLE 11.1

Comparison of Traditional and APT Attacks

	Traditional Attacks	APT Attacks
Attacker	Mostly single person	Highly organized, sophisticated, determined, and well-resourced group
Target	Unspecified, mostly individual systems	Specific organizations, governmental institutions, commercial enterprises
Purpose	Financial benefits, demonstrating abilities	Competitive advantages, strategic benefits
Approach	Single-run, "smash and grab", short period	Repeated attempts, stays low and slow, adapts to resist defenses, long term

Attack Trees as shown in Figure 11.4. The information and assumptions of a security system, possible ways of performing attacks, and various stages of attack can be described using an Attack Tree. Giura et al. [99] have presented APT models using Attack Tree. The authors presented sub-trees as different attack planes and a pointed correlation between different planes. Attack trees can also help the defender in deciding the security strategy for placement of security monitoring and intrusion detection tools. Another key benefit of using an attack tree is that the defender can use alert correlation from various

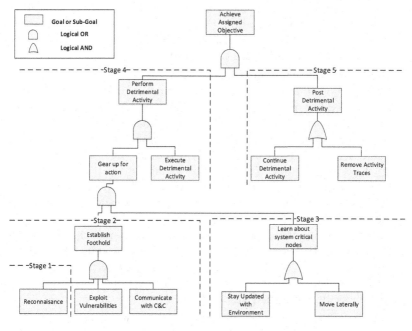

FIGURE 11.4
Stages of advanced persistent threat (APT).

detection mechanisms to check how far the attacker is from his goal (root node of the attack tree). The rectangular nodes in the figure represent the collection of one or more actions, with the topmost rectangular node in each stage being the goal of that particular stage. Using the threat score value assigned to each of the actions (leaf-nodes) and correlation of alerts generated by security tools, the defender will be able to assess the risk and response that is required to mitigate that threat. The APT attack is usually spread across the stages described below:

1. **Reconnaissance:** The beginning of a successful attack starts with the reconnaissance phase. In this stage, the attacker tries to gather a lot of information about the target. The information can be details of employees of the target organization such as social life, websites visited, habits of the employees, details of underlying IT infrastructure, such as kind of hardware switches, routers, anti-virus tools used, web servers available to the public network, open ports, etc. The information can help attackers to easily establish the foothold into the target network. In the attack tree gathering information involves social engineering, reconnaissance, port scanning, service scanning and psychological manipulation of people into accomplishing the goals of attack. Additionally, APT campaigns also query the public repositories such as WHOIS, BGP looking for the domains and routing information. The attackers try to find the websites with high-risk vulnerabilities, such as Cross-Site Scripting (XSS), SQL Injection (SQLI), access points, virtual hosts, firewalls, and IDS/IPS. The reconnaissance is passive in nature and hence difficult to detect.

2. **Foothold Establishment:** The attackers use the reconnaissance information in order to prepare an attack plan. The information collected can be used to exploit the vulnerabilities found in the target organization's web application, databases, and other software. The details of well-known vulnerabilities can be obtained from vulnerability databases such as Common Vulnerability and Exposures List (CVE), Open Source Vulnerability Database (OSVDB) [209], and NIST National Vulnerability Database (NVD) [190], where the known and disclosed vulnerabilities have been archived with a vulnerability identifier (e.g., CVE-ID). Additionally, attackers can also find useful information about the vulnerabilities from dark-web and deep-web forums [198]. Many organizations do not update the current versions of their software even after vulnerability information has been released to maintain business continuity, which allows attackers to exploit the unpatched vulnerabilities.

   Malware is another way of exploiting the target system. According to Symantec's 2017 Internet Security Threat Report, there were 375M malware variants in the year 2016, with email malware rate showing a significant increase, i.e., from 1 in 220 emails in 2015 to 1 in 131 in 2016. The malware can be sent to the victim via Spear-phishing, USB devices, and/or web downloads. The attackers selectively use Business Email Compromise scams after performing social engineering in the initial phase of the attack. The cleverly crafted emails are opened by unaware employees leading to installation and execution of the malware. These techniques as shown in the attack tree help attacker establish a foothold in the target organization network.

3. **Lateral Movement/Stay Undetected:** Once the attacker has gained a foothold into the target environment, he/she can try to move laterally in the target environment and gain access to other sensitive hosts and critical information in the organizational network. The malware can spread to the neighboring machines in the target environment of the infected system. The goal of this phase is to expand the foothold to other systems in search of the data they want to exfiltrate. Once the attacker has reached this stage, it is difficult to completely push the attacker out of the environment. Some methods used by attackers during this stage of attack include user account Password Dumping, Hash Dumping, etc. The harvested credentials facilitate the lateral movement of the attacker and help him in obtaining the restricted information. Tools such as Windows Credential Editor (WCE), Mimikatz, and Windows Local Security Authority (LSA) process are commonly used for dumping credentials.

4. **Exfiltration/Impediment:** The attacker tries to exfiltrate the data collected to his command and control center. Most of the intrusion detection and prevention systems do ingress filtering and not egress filtering, data-exfiltration often goes undetected.

   The attacker may split the exfiltrated data into batches and distribute exfiltration over a long period of time in order to evade the organization's incident detection tools. Table 11.2 shows the most noticeable data exfiltration incidents reported in the year 2017.

5. **Post-Exfiltration/Post-Impediment:** The final phase of the APT is to maintain the persistence until the attack has been lifted by the attack sponsor. The sponsor can choose to lift the attack once the desired goal has been achieved or keep it going to get data from the target continuously. Another important step in this phase for the attacker is to cover his/her tracks so that the attack detection mechanisms cannot trace the attack sponsor or the attacker using system activity and logs.

11.2.3 APT Case Studies

In this sub-section, we discuss some of the notable APTs that affected healthcare, government organizations, and critical Supervisory Control and Data Acquisition (SCADA) systems.

TABLE 11.2

Most Noticeable Data Exfiltration Incidents in 2017

Date	Organization	Number of Affected People	What Got Leaked
19th June	Republican National Committee	200 million	Names, Phone No., Home addresses, Voting details, DOB
13th July	Verizon	14 million	Names, Phone No., PINs
15th Mar	Dun & Bradstreet (DB)	33.7 million	Email addresses, Contract information
12th Mar	Kansas Department of Commerce	5.5 million	Social security numbers
21st Mar	America's Job Link Alliance (AJLA)	4.8 million	Names, DOB, Social security numbers
7th July	World Wrestling Entertainment (WWE)	3 million	Names, Earnings, Ethnicity, Address, Age range
17th July	DOW JONES	2.2 million	Names, Customer IDs, Email addresses
29th July	Equifax	143 million	Social security numbers, Names, Addresses, Driver's Licenses
1st Aug	E-Sports Entertainment Association (ESEA)	1.5 million	Locations, Login details, Email addresses, DOB, Phone No.

11.2.3.1 Stuxnet

Stuxnet is a malware that was discovered in 2010; it targets SCADA systems and is believed to have caused substantial damage to Iran's nuclear program. Stuxnet specifically targeted programmable logic controllers (PLCs), which are used for automation of machinery on factory assembly lines, adventure parks, and nuclear facility centrifuges. The malware used by Stuxnet targeted Microsoft Windows OS, network software, and Siemens Step7 software. Stuxnet collected critical information from the PLCs situated in Iranian nuclear facility and caused a fast spinning of PLC controlled centrifuges, causing centrifuges to tear themselves apart. The worm affected 200,000 computers in the nuclear plant, causing 1000 machines to physically degrade.

Stuxnet consists of three modules: a worm that executes the routines related to the main attack payload, a link file that auto-executes the propagated copies of the worm, and a rootkit component that is responsible for hiding the activities of malicious processes and files, thus preventing the discovery of the worm. The initial delivery mechanism for the worm was using a USB drive. The worm performs the reconnaissance on the network with the goal of discovering Siemens Step7 software - *Step 1 of the APT attack tree*. In the later stages of an attack, worm introduces the infected rootkit on to PLC and Step7 software. The rootkit modifies the code of the software and gives unexpected commands to the PLC while returning loop of normal system operation values to the system users.

11.2.3.2 Hydraq

Hydraq is another example of sophisticated APT that targets highly valued corporate networks. It is also referred to as Operation Aurora, Google Hack Attack and Microsoft Internet Explorer 0-day (CVE-2010-0249). The steps utilized in Hydraq APT have been described below.

1. **Reconnaisance:** The tools such as Whois, DNS information and IP/Network scan could provide initial information about the infrastructure of the target organization. Additionally, attackers utilize social engineering to learn more about the attack target. The target profiling performed by attackers also includes maintaining information about employees, visitors, and contractors who have knowledge and access to the target organization.

2. **Zero day Hack Attack:** Hydraq exploits zero day vulnerability once it is able to discover the outdated version of Microsoft windows explorer on the target. The attackers found an opportunity to target Internet Explorer (IE) by tricking users into visiting a compromised website. The vulnerability *CVE-2010-049/MS10-002* obfuscates the JavaScript code to conceal real intention of the attacker. The code takes advantage of an HTML object handling flaw present in IE when IE tries to access deleted or un-initialized object.

3. **Code Execution:** Once the exploit is successful, the Hydraq's shellcode executes on the target system. Simple bitwise operations such as *XOR* with hardcoded keys such as 0xD8 reveals hidden instruction present in the obfuscated code and Win32/Hydraq installer location.

4. **Maintaining Persistence:** The attackers use backdoors to connect to command and control center and maintain access. Hydraq dropper is responsible for the installation of a dynamically linked library (DLL) component, which contains features and functionality for the remote attacker. Upon execution, dropper generates a random service and makes corresponding entries in the registry file system of Windows. It then creates and starts service under the context of host process *Svchost.exe*.

5. **Deleting Traces:** Hydraq dropper, in addition to the installation of DLL, also clears the traces on the target system to avoid forensic analysis. Additionally, dropper creates and executes a batch file, e.g., *DFS.bat* to remove the dropper file from the system once the end goal of the attack has been achieved.

11.2.4 APT Detection/Mitigation

There have been several efforts towards the development of strategies to cope with targeted attacks like APTs. Many organizations have issued guidelines for dealing with social engineering techniques such as continuous information security training of the employees. Gartner [166] recommends 1) upgrading network and perimeter security, and 2) incorporating strategies to protect against malicious content.

A summary of protection techniques employed by various market leaders in the field of security for APTs has been discussed in the report by Radicati Group [183]. Symantec, for instance, employs on-premises, hybrid and cloud-based solutions for dealing with APT scenarios. Symantec's Advanced Threat Protection (ATP) module comprises of various sub-modules for network, endpoint, email, and roaming (protection against web-based attacks outside corporate network). Force point utilized advanced data loss prevention (DLP), malware protection, insider threat detection tools, and Next-Generation Firewall (NGFW) to detect and mitigate APTs. Other basic prevention techniques such as application whitelisting, patching of software and vulnerabilities, restricting administrative privileges to OS, and applications based on user duties have been used for APT defense in both academic research and industry.

We classify APT detection/mitigation methods into: 1) Monitoring Methods, 2) Detection Methods, and 3) Mitigation Methods.

11.2.5 Orchestrating SDN to Disrupt APT

The design of SDN can help in creating a defense against APT. The SDN architecture having centralized command and network-wide visibility can be extended to include microsegmentation and service chaining-based control in order to break the lateral movement of the attacker during the third phase of the APT attack.

Consider Figure 11.5, that attacker exploits a vulnerability on the web server present on application server A and uses the elevated privileges to exploit the communication server present on the application server B. Similarly, other applications present on the adjacent networks can be targeted by the attacker in a multi-stage attack.

11.2.5.1 SDN-based MicroSegmentation

The controller can centrally enforce microsegmentation policy, for instance, WEBA can communicate with FEECOM002 and CRMAPP2, similarly, the applications within the same sub-network as in Figure 11.6 are allowed to communicate with each other, all other traffic is blocked by default as a part of the microsegmentation policy. Thus the lateral movement of the attacker is localized only to the infected host/application. The microsegmentation can be applied at various levels using SDN controller, e.g., a network gateway,

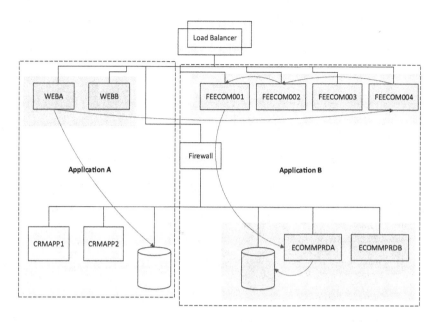

FIGURE 11.5
APT lateral movement example.

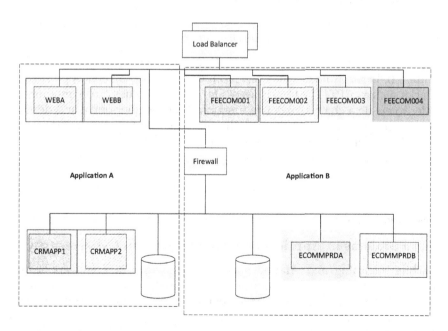

FIGURE 11.6
Micro-segmented cloud architecture to prevent lateral movement of the attacker.

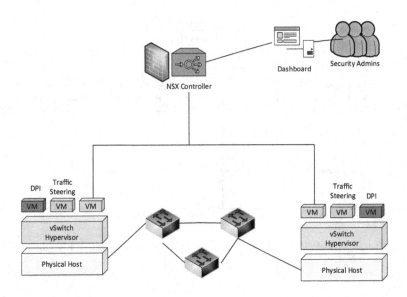

FIGURE 11.7
Defense-in-depth using secured service function chaining.

subnet level, host firewall level, in effect creating a distributed security frame-work managed by a centralized controller.

11.2.5.2 SDN-enabled Secured Service Function Chaining

The operation of defense tools such as IDS/IPS, firewall, and Data Loss Pre-vention (DLP) in isolation may have limited protection against sophisticated attacks such as APT. The flow rules of OpenFlow switches can be modified in order to create a chain of security functions between the source and destina-tion of the network traffic. Using the SDN architecture the traffic can be steered through a series of inspection mechanisms as shown in Figure 11.7, thus increasing the likelihood of APT attack detection/mitigation.

11.3 Problems in Application of Intelligence in Cybersecurity

Machine Learning and AI find successful use cases in many application domains such as recommendation systems, spam detection, speech and image recognition. However, the application of intelligence in cybersecurity, in gene-ral, can be quite challenging. The Network Intrusion Detection Systems (NIDS) that utilize *misuse detection* and *anomaly detection* to flag the abnormal user behavior as an intrusion event can suffer from false positives and false negatives. The cost of misclassifying the normal user activity such as *failed login* attempts as abnormal, and failure to identify malicious activity correctly can prove to be quite costly.

Sommer et al. [255] identify some characteristics exhibited by application of machine learning in NIDS. The characteristics of NIDS are not well aligned with machine learning requirements such as i) High cost of errors; ii) lack of suitable training data; iii) semantic gap between results predicted by ML and their operational interpretation; and iv) high variation in the input data. The sound evaluation of results predicted by the ML algorithm requires domain expertise and semantic *insights* into systems capabilities rather than treating the system as a *black box*.

The ML and AI systems face some generic problems such as false positive and false negative. The cybersecurity domain enhances the probability of errors because adversarial users can try to evade detection by an anomaly-based NIDS, by teaching the system to accept malicious activity as benign. We discuss some issues that can limit the application of intelligence in this section with NIDS as a use case.

11.3.1 Outlier Detection

Machine Learning performs well in classification problems (categorizing similar items) as opposed to detection of anomalies. The classification algorithms utilize *collaborative filtering* to match user preferences and positive ratings for provided items to recommend similar items. Anomaly detection algorithm, on the other hand, would try to identify an anomalous pair of items, i.e., products with no common customers. The machine learning typically has a training phase followed by the testing phase. The training phase would require a large number of samples of both normal as well as abnormal activity in case of NIDS. In a real-world setting, most of the traffic is normal, thus ML-based NIDS systems end up training the detection algorithm on only one class of training samples. The closed world assumption that any test sample not matching the feature set of normal traffic is anomalous is not practical for the real world. Thus, one needs to train an ML algorithm on normal as well as attack traffic to successfully apply ML techniques in the field of NIDS.

11.3.2 High Cost of Errors

The cost of misclassification is much higher for an organization in case of NIDS as opposed to using cases such as image recognition and recommendation systems. For instance, a false positive in the case of a network intrusion event can lead to loss of service for a normal user and wastes significant man-hours for an analyst who is responsible for analyzing the intrusion activity. Even a few false positives can render the NIDS useless [33]. False negatives can, on the other hand, disrupt the security of the organization significantly.

The spam detection and recommendation systems, on the other hand, can tolerate a small fraction of errors.

11.3.3 Semantic Gap

Anomaly detection systems suffer from the problem of the semantic gap between actionable reports based on the output of NIDS and the semantic meaning of the reports from the network operator's point of view. The difference between abnormal behavior and malicious activity should be identified in order to reliably predict network attacks using machine learning. In order to make a production-grade NIDS, the semantic gap issue should be addressed. Additionally, local security-policies and site-specific properties should be identified and incorporated in definition of *malicious* vs. *benign* activity in intrusion detection. For example, *peer-to-peer (P2P)* traffic is considered normal as part of site-specific properties in a network, whereas the absence of this information in training phase of NIDS can cause P2P traffic being flagged as malicious.

11.3.4 Variance in Network Traffic

The features of network traffic such as bandwidth, latency, and network protocols can show significant variation even within the same network environment. Data transfer between applications within the same environment can show a spike in network traffic. An anomaly detection system can find such variability difficult to interpret when distinguishing between normal and abnormal activities. Some operational knowledge of the network is required in such cases so that the traffic can be aggregated over hours, days and weeks. Such traffic shows a reliable pattern for performing intrusion detection. Traffic diversity can also show variance because of application-specific features.

Summary

Intelligent Software-Defined Security discussed in this chapter consists of, but is not limited to, the application of machine learning (ML) and artificial intelligence (AI) in the field of cybersecurity. We discuss different aspects of ML and AI such as search, attack behavior learning, and constraint solving that can help in the detection of network security attacks. The chapter considers an intrusion detection system (IDS) as a specific use case and shows how AI can help in data-reduction and an attacker's behavior classification for an IDS. The role SDN can play in provisioning on an intelligent software-defined security (SDS) has been discussed in the chapter, with DoS protection, and topology protection examples. Traditional attacks have been compared to new emerging threat models such as advanced persistent threats (APTs) in the second section of this chapter. The APT attack model, famous use cases such as Hydraq and Stuxnet, and SDN-based APT detection and mitigation techniques such as microsegmentation and secured SFC are described in this section. The intelligent ML- and AI-based techniques, while they sound quite attractive, suffer from some key limitations, such as high cost of errors, and problems associated with outlier detection in the field of cybersecurity. These limitations have been highlighted in the last section of this chapter. Some important topics that come under the scope of advanced SDN Security such as protection of edge-cloud, vehicular networks [173], and mobile/wireless SDN security [61] are beyond the scope of this book.

Bibliography

1. Understanding and Configuring VLANs. Available at http://www.cisco.com/c/en/us/td/docs/switches/lan/catalyst4500/12-2/25ew/configuration/guide/conf/vlans.html.
2. Apache Karaf. http://karaf.apache.org/.
3. CloudStack. https://cloudstack.apache.org/.
4. DragonFlow. https://wiki.openstack.org/wiki/Dragonflow.
5. Floodlight. http://www.projectfloodlight.org/floodlight/.
6. IEEE 802. 5 Web Site. Available at http://www.ieee802.org/5/www8025org/, retrieved Feb 2018.
7. The Linux Foundation. https://www.linuxfoundation.org/.
8. Open Network Foundation. https://www.opennetworking.org/sdn-resources/openflow.
9. Open vSwitch. http://openvswitch.org/.
10. OpenStack Foundation. https://www.openstack.org/foundation/.
11. OPNFV. https://www.opnfv.org/.
12. Wire Speed to PPS. https://kb.juniper.net/InfoCenter/index?page=content&id=KB14737.
13. Media Access Control (MAC) Bridges. Technical report, 2004.
14. OpenFlow Switch Specification v1.3.1. Technical report, Open Networking Foundation, September 2012.
15. ITU Releases 2014 ICT Figures. https://www.itu.int/en/ITU-D/Statistics/Documents/facts/ICTFactsFigures2014-e.pdf, 2014. Accessed: 2015-09-11.
16. OpenDaylight Project Repository. https://github.com/opendaylight/l2switch, May 2014.
17. Ahmed AbdelSalam, Francois Clad, Clarence Filsfils, Stefano Salsano, Giuseppe Siracusano, and Luca Veltri. Implementation of Virtual Network Function Chaining Through Segment Routing in a Linux-Based nfv Infrastructure. In *Network Softwarization (NetSoft), 2017 IEEE Conference on*, pages 1–5. IEEE, 2017.
18. Ijaz Ahmad, Suneth Namal, Mika Ylianttila, and Andrei Gurtov. Security in Software Defined Networks: A Survey. *IEEE Communications Surveys & Tutorials*, 17(4):2317–2346, 2015.
19. Ian F Akyildiz, Ahyoung Lee, Pu Wang, Min Luo, and Wu Chou. A Roadmap for Traffic Engineering in SDN-Openflow Networks. *Computer Networks*, 71:1–30, October 2014.
20. Mohammad Al-Fares, Sivasankar Radhakrishnan, Barath Raghavan, Nelson Huang, and Amin Vahdat. Hedera: Dynamic Flow Scheduling for Data Center Networks. In *Proceedings of the 7th USENIX Symposium on Networked Systems Design and Implementation (NSDI '10)*, volume 10, pages 19–19. USENIX Association, 2010.

21. Ehab Al-Shaer and Saeed Al-Haj. Flowchecker: Configuration Analysis and Verification of Federated Openflow Infrastructures. In *Proceedings of the 3rd ACM Workshop on Assurable and Usable Security Configuration (SafeConfig '10)*, pages 37–44. ACM, 2010.

22. Ehab Al-Shaer, Hazem Hamed, Raouf Boutaba, and Masum Hasan. Conflict Classification and Analysis of Distributed Firewall Policies. *IEEE Journal on Selected Areas in Communications*, 23(10):2069–2084, 2005.

23. Ehab S Al-Shaer and Hazem H Hamed. Firewall Policy Advisor for Anomaly Discovery and Rule Editing. In *Proceedings of the 8th IFIP/IEEE International Symposium on Integrated Network Management (IM 2003)*, pages 17–30. IEEE, 2003.

24. Joaquin Garcia Alfaro, Nora Boulahia-Cuppens, and Fredric Cuppens. Complete Analysis of Configuration Rules to Guarantee Reliable Network Security Policies. *International Journal of Information Security*, 7(2):103–122, April 2008.

25. Abdullah Alshalan, Sandeep Pisharody, and Dijiang Huang. A Survey of Mobile VPN Technologies. *IEEE Communications Surveys & Tutorials*, 18 (2):1177–1196, 2016.

26. Adel Alshamrani, Ankur Chowdhary, Sandeep Pisharody, Duo Lu, and Dijiang Huang. A Defense System for Defeating DDoS Attacks in SDN Based Networks. In *Proceedings of International Symposium on Mobility Management and Wireless Access (MobiWac)*. IEEE, 2017.

27. Izzat Alsmadi and Dianxiang Xu. Security of Software Defined Networks: A Survey. *Computers & Security*, 53:79–108, 2015.

28. Paul Ammann, Joseph Pamula, Ronald Ritchey, and Julie Street. A Host-Based Approach to Network Attack Chaining Analysis. In *Computer Security Applications Conference, 21st Annual*, pages 10-pp. IEEE, 2005.

29. Paul Ammann, Duminda Wijesekera, and Saket Kaushik. Scalable, Graph-Based Network Vulnerability Analysis. In *Proceedings of the 9th ACM Conference on Computer and Communications Security*, pages 217–224. ACM, 2002.

30. David G Andersen. Theoretical Approaches to Node Assignment. *Computer Science Department*, page 86, 2002.

31. Apache Groovy. Available at http://groovy-lang.org/.

32. Daniel Awduche, Angela Chiu, Anwar Elwalid, Indra Widjaja, and XiPeng Xiao. Overview and Principles of Internet Traffic Engineering. RFC 3272, IETF, 2002.

33. Stefan Axelsson. The Base-Rate Fallacy and its Implications for the Difficulty of Intrusion Detection. In *Proceedings of the 6th ACM Conference on Computer and Communications Security*, pages 1–7. ACM, 1999.

34. Jie Bai, Yu Wu, Guoyin Wang, Simon X Yang, and Wenbin Qiu. A Novel Intrusion Detection Model Based on Multi-Layer Self-Organizing Maps and Principal Component Analysis. In *International Symposium on Neural Networks*, pages 255–260. Springer, 2006.

35. James Bailey, George Papamarkos, Alexandra Poulovassilis, and Peter T Wood. An Event-Condition-Action Language for XML. In *Web Dynamics*, pages 223–248. Springer, 2004.

36. Kapil Bakshi. Considerations for Software-Defined Networking (SDN): Approaches and Use Cases. In *Proceedings of the 2013 IEEE Aerospace Conference*, pages 1–9. IEEE, 2013.

37. Md Faizul Bari, Arup Raton Roy, Shihabur Rahman Chowdhury, Qi Zhang, Mohamed Faten Zhani, Reaz Ahmed, and Raouf Boutaba. Dynamic Controller Provisioning in Software-Defined Networks. In *Proceedings of the 9th International Conference on Network and Service Management (CNSM 2013)*, pages 18–25. IEEE, 2013.

38. Yair Bartal, Alain Mayer, Kobbi Nissim, and Avishai Wool. Firmato: A Novel Firewall Management Toolkit. In *Proceedings of the 1999 IEEE Symposium on Security and Privacy*, pages 17–31. IEEE, 1999.

39. Cataldo Basile, Alberto Cappadonia, and Antonio Lioy. Algebraic Models to Detect and Solve Policy Conflicts. In *Proceedings of the 7th International Conference on Mathematical Methods, Models, and Architectures for Computer Network Security (MMM-ACNS 2007)*, pages 242–247. Springer, 2007.

40. Beheshti, N. and Zhang, Y. Fast Failover for Control Traffic in Software-Defined Networks. In *Proceedings of the 2012 IEEE Global Communications Conference (GLOBECOM 2012)*, pages 2665–2670. IEEE, December 2012.

41. Steven M Bellovin. Distributed Firewalls. http://static.usenix.org/publications/login/1999-11/features/firewalls.html, 1999.

42. Theophilus Benson, Aditya Akella, and David A Maltz. Network Traffic Characteristics of Data Centers in the Wild. In *Proceedings of the 10th A CM SIG-COMM Conference on Internet Measurement (IMC 10)*, pages 267–280. ACM, 2010.

43. Theophilus Benson, Ashok Anand, Aditya Akella, and Ming Zhang. MicroTE: Fine Grained Traffic Engineering for Data Centers. In *Proceedings of the Seventh Conference on Emerging Networking Experiments and Technologies (CoNEXT '11)*, page 8. ACM, 2011.

44. Jon Louis Bentley. Multidimensional Binary Search Trees Used for Associative Searching. *Communications of the ACM*, 18(9):509–517, 1975.

45. Pankaj Berde, Matteo Gerola, Jonathan Hart, Yuta Higuchi, Masayoshi Kobayashi, Toshio Koide, Bob Lantz, Brian O'Connor, Pavlin Radoslavov, William Snow, and others. ONOS: Towards an Open, Distributed SDN OS. In *Proceedings of the 3rd Workshop on Hot Topics in Software Defined Networking (HotSDN 2014)*, pages 1–6. ACM, 2014.

46. Deval Bhamare, Raj Jain, Mohammed Samaka, and Aiman Erbad. A Survey on Service Function Chaining. *Journal of Network and Computer Applications*, 75:138–155, 2016.

47. Bierman, A., Bjorklund, M., and Watsen, K. RESTCONF Protocol. RFC 8040, IETF, January 2017.

48. Matt Blaze, Joan Feigenbaum, and Angelos D Keromytis. Keynote: Trust Management for Public-Key Infrastructures. In *International Workshop on Security Protocols*, pages 59–63. Springer, 1998.

49. Tyler Bletsch, Xuxian Jiang, Vince W Freeh, and Zhenkai Liang. Jump-Oriented Programming: A New Class of Code-Reuse Attack. In *Proceedings of the 6th ACM Symposium on Information, Computer and Communications Security*, pages 30–40. ACM, 2011.

50. Pat Bosshart, Dan Daly, Glen Gibb, Martin Izzard, Nick McKeown, Jennifer Rexford, Cole Schlesinger, Dan Talayco, Amin Vahdat, George Varghese, et al. P4: Programming Protocol-Independent Packet Processors. *ACM SIG-COMM Computer Communication Review*, 44(3):87–95, 2014.

51. Pat Bosshart, Dan Daly, Glen Gibb, Martin Izzard, Nick McKeown, Jennifer Rexford, Cole Schlesinger, Dan Talayco, Amin Vahdat, George Varghese, et al. P4 Language Git Repository. https://github.com/p4lang/, 2017.

52. Zdravko Bozakov. Towards Virtual Routers as a Service. In *Proc. of 6th GI/ITG KuVS Workshop on Future Internet*, 2010.

53. Levente Buttyn, Gbor Pk, and Ta Vinh Thong. Consistency Verification of Stateful Firewalls Is Not Harder Than the Stateless Case. *Infocommunications Journal*, 64(1):2–8, 2009.

54. Filipe Caldeira and Edmundo Monteiro. A Policy-Based Approach to Firewall Management. In *Proceedings of the IFIP TC6 / WG6.2 & WG6.7 Conference on Network Control and Engineering for QoS, Security and Mobility (Net-Con '02)*, pages 115–126. Springer, 2003.

55. Venanzio Capretta, Bernard Stepien, Amy Felty, and Stan Matwin. Formal Correctness of Conflict Detection for Firewalls. In *Proceedings of the 2007 ACM Workshop on Formal Methods in Security Engineering (FMSE '07)*, pages 22–30. ACM, 2007.

56. Ramaswamy Chandramouli and Ramaswamy Chandramouli. Secure Virtual Network Configuration for Virtual Machine (vm) Protection. *NIST Special Publication 800–125B*, 2016.

57. V Chatzigiannakis, G Androulidakis, and B Maglaris. A Distributed Intrusion Detection Prototype Using Security Agents. *HP OpenView University Association*, 2004.

58. Yi Cheng, Julia Deng, Jason Li, Scott A DeLoach, Anoop Singhal, and Xinming Ou. Metrics of security. In *Cyber Defense and Situational Awareness*, pages 263–295. Springer, 2014.

59. Margaret Chiosi, Don Clarke, Peter Willis, Andy Reid, James Feger, Michael Bugenhagen, Waqar Khan, Michael Fargano, Chunfeng Cui, Hui Deng, et al. Network Functions Virtualisation: An Introduction, Benefits, Enablers, Challenges and Call for Action. In *SDN and Open-Flow World Congress*, pages 22–24, 2012.

60. Laurence Cholvy and Frédéric Cuppens. Analyzing Consistency of Security Policies. In *Proceedings of the 1997 IEEE Symposium on Security and Privacy*, pages 103–112. IEEE, 1997.

61. Ankur Chowdhary. *Secure Mobile SDN*. Arizona State University, 2015.

62. Ankur Chowdhary, Vaibhav Hemant Dixit, Naveen Tiwari, Sukhwa Kyung, Dijiang Huang, and Gail-Joon Ahn. Science DMZ: SDN Based Secured Cloud Testbed. In *Network Function Virtualization and Software Defined Networks (NFV-SDN), 2017 IEEE Conference on*, pages 1–2. IEEE, 2017.

63. Ankur Chowdhary, Sandeep Pisharody, Adel Alshamrani, and Dijiang Huang. Dynamic Game Based Security Framework in SDN-Enabled Cloud Networking Environments. In *Proceedings of the ACM International Workshop on Security in Software Defined Networks & Network Function Virtualization*, pages 53–58. ACM, 2017.

64. Ankur Chowdhary, Sandeep Pisharody, and Dijiang Huang. SDN Based Scalable MTD Solution in Cloud Network. In *Proceedings of the 2016 ACM Workshop on Moving Target Defense*, pages 27–36. ACM, 2016.

65. Omar Chowdhury, Andreas Gampe, Jianwei Niu, Jeffery von Ronne, Jared Bennatt, Anupam Datta, Limin Jia, and William H Winsborough. Privacy Promises That Can Be Kept: A Policy Analysis Method with

Application to the HIPAA Privacy Rule. In *Proceedings of the 18th ACM Symposium on Access Control Models and Technologies (SACMAT '13)*, pages 3–14. ACM, 2013.

66. Andrew Clark, Kun Sun, Linda Bushnell, and Radha Poovendran. A Game-Theoretic Approach to IP Address Randomization in Decoy-Based Cyber Defense. In *International Conference on Decision and Game Theory for Security*, pages 3–21. Springer, 2015.

67. Amazon Elastic Compute Cloud. Amazon Web Services. *Retrieved November*, 9:2011, 2011.

68. Warren Connell, Daniel A Menascé, and Massimiliano Albanese. Performance Modeling of Moving Target Defenses. In *Proceedings of the 2017 Workshop on Moving Target Defense*, pages 53–63. ACM, 2017.

69. Michelle Cotton, Lars Eggert, Joe Touch, Magnus Westerlund, and Stuart Cheshire. Internet Assigned Numbers Authority (IANA) Procedures for the Management of the Service Name and Transport Protocol Port Number Registry. Technical report, 2011.

70. Andrew R Curtis, Wonho Kim, and Praveen Yalagandula. Mahout: Low-Overhead Datacenter Traffic Management Using End-Host-Based Elephant Detection. In *Proceedings of the 30th International IEEE Conference on Computer Communications (INFOCOM 2011)*, pages 1629–1637. IEEE, 2011.

71. Andrew R Curtis, Jeffrey C Mogul, Jean Tourrilhes, Praveen Yalagandula, Puneet Sharma, and Sujata Banerjee. DevoFlow: Scaling Flow Management for High-Performance Networks. *ACM SIGCOMM Computer Communication Review*, 41:254–265, 2011.

72. CVSS. Common Vulnerability Scoring System (CVSS). https://www.first.org/cvss/, 2017. [Online; accessed 01-15-2018].

73. Frank D'Agostino. ACI Surpasses VMware NSX Again with Micro Segmentation and End-Point Granularity. *Blog at Cisco*, 2016.

74. Frank D'Agostino. Cisco Application Centric Infrastructure (aci). https://www.cisco.com/c/en/us/solutions/data-center-virtualization/application-centric-infrastructure/index.html, 2018.

75. Nicodemos C Damianou. *A Policy Framework for Management of Distributed Systems*. PhD Dissertation, Imperial College, 2002.

76. Huynh Tu Dang, Han Wang, Theo Jepsen, Gordon Brebner, Changhoon Kim, Jennifer Rexford, Robert Soulé, and Hakim Weatherspoon. Whip-Persnapper: A p4 Language Benchmark Suite. In *Proceedings of the Symposium on SDN Research*, pages 95–101. ACM, 2017.

77. Saurav Das, Guru Parulkar, Nick McKeown, Preeti Singh, Daniel Getachew, and Lyndon Ong. Packet and Circuit Network Convergence with Openflow. In *Optical Fiber Communication Conference*, page OTuG1. Optical Society of America, 2010.

78. Derek L Davis. Secure boot, August 10 1999. US Patent 5,937,063.

79. Saptarshi Debroy, Prasad Calyam, Minh Nguyen, Allen Stage, and Vladimir Georgiev. Frequency-Minimal Moving Target Defense Using Software-Defined Networking. In *Computing, Networking and Communications (ICNC), 2016 International Conference on*, pages 1–6. IEEE, 2016.

80. Mohan Dhawan, Rishabh Poddar, Kshiteej Mahajan, and Vijay Mann. SPHINX: Detecting Security Attacks in Software-Defined Networks.

In *Proceedings of the Network and Distributed System Security Symposium 2015 (NDSS 15)*. ISOC, 2015.

81. Advait Dixit, Fang Hao, Sarit Mukherjee, TV Lakshman, and Ramana Kompella. Towards an Elastic Distributed SDN Controller. *ACM SIGCOMM Computer Communication Review*, 43:7–12, 2013.

82. Avri Doria, J Hadi Salim, Robert Haas, Horzmud Khosravi, Weiming Wang, Ligang Dong, Ram Gopal, and Joel Halpern. Forwarding and Control Element Separation (forces) Protocol Specification. Technical report, 2010.

83. Open vSwitch with DPDK. http://docs.openvswitch.org/en/latest/intro/install/dpdk/.

84. Nicodemos Damianou and Naranker Dulay. The Ponder Policy Specification Language. In *Lecture Notes in Computer Science*, pages 18–38. Springer-Verlag, 2001.

85. Iman El Mir, El Mehdi Kandoussi, Mohamed Hanini, Abdelkrim Haqiq, and Dong Seong Kim. A Game Theoretic Approach Based Virtual Machine Migration for Cloud Environment Security. *International Journal of Communication Networks and Information Security*, 9(3):345–357, 2017.

86. David Eppstein and S Muthukrishnan. Internet Packet Filter Management and Rectangle Geometry. In *Proceedings of the 12th Annual ACM-SIAM Symposium on Discrete Algorithms (SODA '01)*, pages 827–835. Society for Industrial and Applied Mathematics, 2001.

87. David Erickson. The Beacon Openflow Controller. In *Proceedings of the 2nd Workshop on Hot Topics in Software Defined Networking (HotSDN 2013)*, pages 13–18. ACM, 2013.

88. ETSI. NFV Security and Trust Guidance, Dec, 2014.

89. Seyed Kaveh Fayazbakhsh, Luis Chiang, Vyas Sekar, Minlan Yu, and Jeffrey C Mogul. Enforcing Network-Wide Policies in the Presence of Dynamic Middlebox Actions Using Flowtags. In *NSDI*, volume 14, pages 533–546, 2014.

90. Nick Feamster, Hari Balakrishnan, Jennifer Rexford, Aman Shaikh, and Jacobus Van Der Merwe. The Case for Separating Routing from Routers. In *Proceedings of the ACM SIGCOMM workshop on Future directions in network architecture*, pages 5–12. ACM, 2004.

91. Andreas Fischer, Juan Felipe Botero, Michael Till Beck, Hermann De Meer, and Xavier Hesselbach. Virtual Network Embedding: A Survey. *IEEE Communications Surveys & Tutorials*, 15(4):1888–1906, 2013.

92. Nate Foster, Rob Harrison, Michael J Freedman, Christopher Monsanto, Jennifer Rexford, Alec Story, and David Walker. Frenetic: A Network Programming Language. In *Proceedings of the 16th ACM SIGPLAN International Conference on Functional Programming (ICFP '11)*, volume 46, pages 279–291. ACM, 2011.

93. Jeremy Frank. Artificial Intelligence and Intrusion Detection: Current and Future Directions. In *Proceedings of the 17th national computer security conference*, volume 10, pages 1–12. Baltimore, MD, 1994.

94. Jérôme François, Lautaro Dolberg, Olivier Festor, and Thomas Engel. Network Security Through Software Defined Networking: A Survey. In *Proceedings of the 7th Conference on Principles, Systems and Applications of IP Telecommunications (IPTComm 2014)*, page 6. ACM, 2014.

95. Zhi Fu, S Felix Wu, He Huang, Kung Loh, Fengmin Gong, Ilia Baldine, and Chong Xu. IPSec/VPN Security Policy: Correctness, Conflict Detection, and

Resolution. In *Proceedings of the International Workshop on Policies for Distributed Systems and Networks (POLICY 2001), volume 1995 of Lecture Notes in Computer Science*, pages 39–56. Springer, 2001.

96. Shang Gao, Zecheng Li, Bin Xiao, and Guiyi Wei. Security Threats in the Data Plane of Software-Defined Networks. *IEEE Network*, 2018.

97. Miguel Garcia, Alysson Bessani, Ilir Gashi, Nuno Neves, and Rafael Obelheiro. Os Diversity for Intrusion Tolerance: Myth or Reality? In *Dependable Systems & Networks (DSN), 2011 IEEE/IFIP 41st International Conference on*, pages 383–394. IEEE, 2011.

98. Aaron Gember, Prathmesh Prabhu, Zainab Ghadiyali, and Aditya Akella. Toward Software-Defined Middlebox Networking. In *Proceedings of the 11th ACM Workshop on Hot Topics in Networks (HotNets-XI)*, pages 7–12. ACM, 2012.

99. Paul Giura and Wei Wang. A Context-Based Detection Framework for Advanced Persistent Threats. In *Cyber Security (CyberSecurity), 2012 International Conference on*, pages 69–74. IEEE, 2012.

100. John C Gower and Gavin JS Ross. Minimum Spanning Trees and Single Link-Age Cluster Analysis. *Applied statistics*, pages 54–64, 1969.

101. Marc Green, Douglas C MacFarland, Doran R Smestad, and Craig A Shue. Characterizing Network-Based Moving Target Defenses. In *Proceedings of the Second ACM Workshop on Moving Target Defense*, pages 31–35. ACM, 2015.

102. Yu Gu, Andrew McCallum, and Don Towsley. Detecting Anomalies in Network Traffic Using Maximum Entropy Estimation. In *Proceedings of the 5th ACM SIGCOMM Conference on Internet Measurement (IMC 05)*, page 32. USENIX Association, 2005.

103. Joshua D Guttman and Amy L Herzog. Rigorous Automated Network Security Management. *International Journal of Information Security*, 4(1-2):29–48, February 2005.

104. Martin T Hagan, Howard B Demuth, Mark H Beale, et al. *Neural network design, volume 20*. Pws Pub. *Boston*, 1996.

105. Joel Halpern and Carlos Pignataro. Service Function Chaining (sfc) Architecture. Technical report, 2015.

106. Bo Han, Vijay Gopalakrishnan, Lusheng Ji, and Seungjoon Lee. Network Function Virtualization: Challenges and Opportunities for Innovations. *IEEE Communications Magazine*, 53(2):90–97, 2015.

107. Weili Han and Chang Lei. A Survey on Policy Languages in Network and Security Management. *Computer Networks*, 56(1):477–489, January 2012.

108. Adiseshu Hari, Subhash Suri, and Guru Parulkar. Detecting and Resolving Packet Filter Conflicts. In *Proceedings of the 19th Annual Joint Conference of the IEEE Computer and Communications Societies (IN-FOCOM 2000), volume 3*, pages 1203–1212. IEEE, 2000.

109. Scott Hazelhurst. Algorithms for Analysing Firewall and Router Access Lists. *CoRR*, cs.NI/0008006, 2000.

110. Charles L Hedrick. Routing Information Protocol. *Technical report*, 1988.

111. Alvaro Herrero, Emilio Corchado, María A Pellicer, and Ajith Abraham. Hybrid Multi Agent-Neural Network Intrusion Detection with Mobile Visualization. In *Innovations in Hybrid Intelligent Systems*, pages 320–328. Springer, 2007.

112. Wade Holmes. Vmware nsx Micro-Segmentation Day 1. VMWare Press, available at https://www.vmware.com/content/dam/digitalmarketing/vmware/en/pdf/products/nsx/vmware-nsx-microsegmentation.pdf, 2018.

113. Ralph Holz, Thomas Riedmaier, Nils Kammenhuber, and Georg Carle. X. 509 Forensics: Detecting and Localising the SSL/TLS Men-in-the-Middle. In *European Symposium on Research in Computer Security*, pages 217–234. Springer, 2012.

114. Homer, John, Xinming Ou, and David Schmidt. A Sound and Practical Approach to Quantifying Security Risk in Enterprise Networks. Kansas State University Technical Report (2009): 1–15.

115. John Homer, Ashok Varikuti, Xinming Ou, and Miles A McQueen. Improving Attack Graph Visualization Through Data Reduction and Attack Grouping. In *Visualization for computer security*, pages 68–79. Springer, 2008.

116. Andrei Homescu, Todd Jackson, Stephen Crane, Stefan Brunthaler, Per Larsen, and Michael Franz. Large-Scale Automated Software Diversityprogram Evolution Redux. *IEEE Transactions on Dependable and Secure Computing*, 14 (2):158–171, 2017.

117. Jin Hong and Dong-Seong Kim. Harms: Hierarchical Attack Representation Models for Network Security Analysis. SRI Security Research Institute, Edith Cowan University, Perth, Western Australia, 2012.

118. Jin B Hong and Dong Seong Kim. Assessing the Effectiveness of Moving Target Defenses Using Security Models. *IEEE Transactions on Dependable and Secure Computing*, 13(2):163–177, 2016.

119. Sungmin Hong, Lei Xu, Haopei Wang, and Guofei Gu. Poisoning Network Visibility in Software-Defined Networks: New Attacks and Counter-Measures. In *NDSS*, volume 15, pages 8–11, 2015.

120. Hongxin Hu, Gail-Joon Ahn, and Ketan Kulkarni. Fame: A Firewall Anomaly Management Environment. In *Proceedings of the 3rd ACM Workshop on Assurable and Usable Security Configuration (SafeConfig '10)*, pages 17–26. ACM, 2010.

121. Hongxin Hu, Wonkyu Han, Gail-Joon Ahn, and Ziming Zhao. Flow-Guard: Building Robust Firewalls for Software-Defined Networks. In *Proceedings of the 3rd Workshop on Hot Topics in Software Defined Networking (HotSDN 2014)*, pages 97–102. ACM, 2014.

122. Yannan Hu, Wendong Wang, Xiangyang Gong, Xirong Que, and Shiduan Cheng. BalanceFlow: Controller Load Balancing for Openflow Networks. In *Proceedings of the 2012 IEEE 2nd International Conference on Cloud Computing and Intelligent Systems (CCIS 2012)*, volume 2, pages 780–785. IEEE, 2012.

123. Yannan Hu, Wendong Wang, Xiangyang Gong, Xirong Que, and Shiduan Cheng. On the Placement of Controllers in Software-Defined Networks. *The Journal of China Universities of Posts and Telecommunications*, 19:92–171, 2012.

124. Yannan Hu, Wendong Wang, Xiangyang Gong, Xirong Que, and Shiduan Cheng. Reliability-Aware Controller Placement for Software-Defined Networks. In *Proceedings of the 13th IFIP/IEEE International Symposium on Integrated Network Management (IM 2013)*, pages 672–675. IEEE, May 2013.

125. Dijiang Huang, Deep Medhi, and Kishor Trivedi. SRN: On Establishing Secure and Resilient Networking Services. Available at https://www.nsf.gov/awardsearch/showAward?AWD_ID=1526299, 2015.

126. Heqing Huang, Su Zhang, Xinming Ou, Atul Prakash, and Karem Sakallah. Distilling Critical Attack Graph Surface Iteratively Through Minimum-Cost Sat Solving. In *Proceedings of the 27th Annual Computer Security Applications Conference*, pages 31–40. ACM, 2011.

127. W. S. Humphrey. Characterizing the Software Process: A Maturity Framework. *IEEE Software*, 5(2):73–79, March 1988.

128. ICC Lab. NETFLOC SDK4SDN. https://blog.zhaw.ch/icclab/service-func-tion-chaining-using-the-sdk4sdn/, 2018. Online; accessed 13 February 2018.

129. IEEE. IEEE Std 802.1Q-1998. IEEE standard, availabe at http://ieeexplore.ieee.org/xpl/RecentIssue.jsp?punumber=6080, 1998.

130. IEEE Standards Association. Guidelines for Use of Extended Unique Identifier (EUI), Organizationally Unique Identifier (OUI), and Company ID (CID). available at http://standards.ieee.org/develop/regauth/tut/eui.pdf, retrieved Feb 2018.

131. Kyle Ingols, Richard Lippmann, and Keith Piwowarski. Practical Attack Graph Generation for Network Defense. In *Computer Security Applications Conference, 2006. ACSAC'06. 22nd Annual*, pages 121–130. IEEE, 2006.

132. Sotiris Ioannidis, Angelos D Keromytis, Steve M Bellovin, and Jonathan M Smith. Implementing a Distributed Firewall. In *Proceedings of the 7th ACM Conference on Computer and Communications Security*, pages 190–199. ACM, 2000.

133. Jafar Haadi Jafarian, Ehab Al-Shaer, and Qi Duan. Openflow Random Host Mutation: Transparent Moving Target Defense Using Software Defined Networking. In *Proceedings of the First Workshop on Hot Topics in Software Defined Networks*, pages 127–132. ACM, 2012.

134. Sushil Jajodia, Steven Noel, and Brian OBerry. Topological Analysis of Network Attack Vulnerability. In *Managing Cyber Threats*, pages 247–266. Springer, 2005.

135. Tariq Javid, Tehseen Riaz, and Asad Rasheed. A Layer2 Firewall for Software Defined Network. In *Proceedings of the 2014 Conference on Information Assurance and Cyber Security (CIACS)*, pages 39–42. IEEE, 2014.

136. Quan Jia, Kun Sun, and Angelos Stavrou. Motag: Moving Target Defense Against Internet Denial of Service Attacks. In *Computer Communications and Networks (ICCCN), 2013 22nd International Conference on*, pages 1–9. IEEE, 2013.

137. Yury Jimenez, Cristina Cervello-Pastor, and Aurelio J Garcia. On the Controller Placement for Designing a Distributed SDN Control Layer. In *Proceedings of the 2014 IFIP Networking Conference (Networking 2014)*, pages 1–9. IEEE, 2014.

138. HyunChul Joh and Yashwant K Malaiya. Defining and Assessing Quantitative Security Risk Measures Using Vulnerability Lifecycle and CVSS Metrics. In *Proceedings of the 10th International Conference on Security and Management (SAM '11)*, pages 10–16, 2011.

139. Postel Jon. Internet Protocol-Darpa Internet Program Protocol Specification. *Technical report*, RFC-791, DARPA, 1981.

140. Dilip A Joseph, Arsalan Tavakoli, and Ion Stoica. A Policy-Aware Switching Layer for Data Centers. In *Proceedings of the 2008 ACM Conference on Special Interest Group on Data Communication (SIGCOMM '08)*. ACM, 2008.

141. Lalana Kagal. Rei: A Policy Language for the Me-Centric Project. *Technical Report*, HP Laboratories, Palo Alto, 2002.

142. Kubra Kalkan, Gurkan Gur, and Fatih Alagoz. Defense Mechanisms Against ddos Attacks in sdn Environment. *IEEE Communications Magazine*, 55(9):175–179, 2017.

143. Panos Kampanakis, Harry Perros, and Tsegereda Beyene. SDN-Based Solutions for Moving Target Defense Network protection. In *World of Wvreless, Mobile and Multimedia Networks (WoWMoM), 2014 IEEE 15th International Symposium on a*, pages 1–6. IEEE, 2014.

144. Hasan T Karaoglu and Murat Yuksel. Offloading Routing Complexity to the Cloud (s). In *Communications Workshops (ICC), 2013 IEEE International Conference on*, pages 1367–1371. Citeseer, 2013.

145. George Karypis and Vipin Kumar. A Parallel Algorithm for Multilevel Graph Partitioning and Sparse Matrix Ordering. *Journal of Parallel and Distributed Computing*, 48(1):71–95, 1998.

146. Seiichi Kawamura and Masanobu Kawashima. A Recommendation for IPV6 Address Text Representation. Technical report, 2010.

147. Ryota Kawashima. vNFC: A Virtual Networking Function Container for SDN-enabled Virtual Networks. In *Proceedings of the 2nd Symposium on Network Cloud Computing and Applications (NCCA 2012)*, pages 124–129. IEEE, 2012.

148. Peyman Kazemian, Michael Chan, Hongyi Zeng, George Varghese, Nick McKeown, and Scott Whyte. Real Time Network Policy Checking Using Header Space Analysis. In *NSDI*, pages 99–111, 2013.

149. Peyman Kazemian, George Varghese, and Nick McKeown. Header Space Analysis: Static Checking for Networks. In *Proceedings of the 9th USENIX Symposium on Networked Systems Design and Implementation (NSDI '12)*, pages 113–126. USENIX Association, 2012.

150. Pankaj Kumar Khatkar. Firewall Rule Set Analysis and Visualization. Master's thesis, Arizona State University, 2014.

151. Gautam Khetrapal and Saurabh Kumar Sharma. Demystifying Routing Services in Software-Defined Networking. *White paper*, 2013.

152. Ahmed Khurshid, Xuan Zou, Wenxuan Zhou, Matthew Caesar, and P Brighten Godfrey. VeriFlow: Verifying Network-Wide Invariants in Real Time. In *Proceedings of the 10th USENIX Symposium on Networked Systems Design and Implementation (NSDI '13)*, pages 15–27. USENIX Association, 2013.

153. Hyojoon Kim and Nick Feamster. Improving Network Management with Software Defined Networking. *IEEE Commun. Mag.*, 51(2):114–119, Feb. 2013.

154. Richard Kissel. *Glossary of Key Information Security Terms*. Diane Publishing, 2011.

155. Jon M Kleinberg. *Approximation Algorithms for Disjoint Paths Problems*. PhD thesis, Massachusetts Institute of Technology, 1996.

156. Knight, S., Nguyen, H. X., Falkner, N., Bowden, R., and Roughan, M. The Internet Topology Zoo. *IEEE Journal on Selected Areas in Communications*, 29(9):1765–1775, October 2011.

157. Ron Kohavi. Scaling up the Accuracy of Naive-Bayes Classifiers: a Decision-Tree Hybrid. In *KDD*, volume 96, pages 202–207. Citeseer, 1996.

158. Kolliopoulos, Stavros G., and Clifford Stein. Improved Approximation Algorithms for Unsplittable Flow Problems. focs. IEEE, 1997.

159. Teemu Koponen, Martin Casado, Natasha Gude, Jeremy Stribling, Leon Poutievski, Min Zhu, Rajiv Ramanathan, Yuichiro Iwata, Hiroaki Inoue, Takayuki Hama, and others. Onix: A Distributed Control Platform for Large-Scale Production Networks. In *Proceedings of the 9th USENIX Symposium on Operating Systems Design and Implementation (OSDI '10)*, volume 10, pages 1–6. USENIX Association, 2010.

160. Maria Korolov and Lysa Myers. What is the Cyber Kill Chain? why it's not always the right approach to cyber attacks. Available at https://www.csoonline.com/, accessed on June 24, 2018.

161. Igor Kotenko, Alexey Konovalov, and Andrey Shorov. Agent-Based Modeling and Simulation of Botnets and Botnet Defense. In *Conference on Cyber Conflict*. CCD COE Publications. Tallinn, Estonia, pages 21–44, 2010.

162. Igor Kotenko and Alexander Ulanov. Multi-Agent Framework for Simulation of Adaptive Cooperative Defense Against Internet Attacks. In *International Workshop on Autonomous Intelligent Systems: Multi-Agents and Data Mining*, pages 212–228. Springer, 2007.

163. Michail-Alexandros Kourtis, Michael J McGrath, Georgios Gardikis, Georgios Xilouris, Vincenzo Riccobene, Panagiotis Papadimitriou, Eleni Trouva, Francesco Liberati, Marco Trubian, Josep Batallé, et al. T-nova: An Open-Source Mano Stack for NFV Infrastructures. *IEEE Transactions on Network and Service Management*, 14(3):586–602, 2017.

164. Diego Kreutz, Fernando MV Ramos, Paulo Esteves Verissimo, Christian Esteve Rothenberg, Siamak Azodolmolky, and Steve Uhlig. Software-Defined Networking: A Comprehensive Survey. *Proceedings of the IEEE*, 103(1):14–76, 2015.

165. Maciej Kuniar, Peter Pereni, and Dejan Kosti. What You Need to Know About SDN Flow Tables. In *International Conference on Passive and Active Network Measurement*, pages 347–359. Springer, 2015.

166. Kaspersky Lab. Strategies for Mitigating Advanced Persistent Threats (APTs). https://securelist.com/threats/strategies-for-mitigating-advanced-persistent-threats-apts/, 2017.

167. Kris Lahiri. How Will Artificial Intelligence And Machine Learning Impact Cyber Security? Available at https://www.forbes.com/, 2018. [Online; accessed 06-04-2018].

168. M. Lasserre and V. Kompella. Virtual Private LAN Service (VPLS) Using Label Distribution Protocol (LDP) Signaling. *RFC4762*, 2007.

169. Jeongkeun Lee, Jean Tourrilhes, Puneet Sharma, and Sujata Banerjee. No More Middlebox: Integrate Processing into Network. *ACM SIGCOMM Computer Communication Review*, 40:459–460, 2010.

170. Richard Paul Lippmann, JF Riordan, TH Yu, and KK Watson. Continuous Security Metrics for Prevalent Network Threats: Introduction and First Four Metrics. Technical Report MIT-LL-IA-3, Massachusetts Institute of Institute of Technology Lincoln Laboratory, May 2012.

171. Jiaqiang Liu, Yong Li, Huandong Wang, Depeng Jin, Li Su, Lieguang Zeng, and Thanos Vasilakos. Leveraging Software-Defined Networking for Security Policy Enforcement. *Information Sciences*, 327:288–299, 2016.

172. Xiong Liu, Haiwei Xue, Xiaoping Feng, and Yiqi Dai. Design of the Multi-Level Security Network Switch System Which Restricts Covert Channel. In *Proceedings of the IEEE 3rd International Conference on Communication Software and Networks (ICCSN 2011)*, pages 233–237. IEEE, 2011.

173. Duo Lu, Zhichao Li, Dijiang Huang, Xianglong Lu, Yuli Deng, Ankur Chowdhary, and Bing Li. VC-bots: A Vehicular Cloud Computing Testbed with Mobile Robots. In *Proceedings of the First International Workshop on Internet of Vehicles and Vehicles of Internet*, pages 31–36. ACM, 2016.

174. Ting Luo and Shaohua Yu. Control and Communication Mechanisms of a Softrouter. In *Optical Internet and Next Generation Network, 2006. COIN-NGNCON 2006. The Joint International Conference on*, pages 109–111. IEEE, 2006.

175. Emil C Lupu and Morris Sloman. Conflict Analysis for Management Policies. In *Integrated Network Management V, IFIP - The International Federation for Information Processing*, pages 430–443. Springer, 1997.

176. Emil C Lupu and Morris Sloman. Conflicts in Policy-Based Distributed Systems Management. *IEEE Transactions on Software Engineering*, 25(6):852–869, 1999.

177. Hugh Mahon, Yoram Bernet, Shai Herzog, and Jhon Schnizlein. Requirements for a Policy Management System. *Internet Draft*, IETF, 1999.

178. Hoda Maleki, Saeed Valizadeh, William Koch, Azer Bestavros, and Marten van Dijk. Markov Modeling of Moving Target Defense Games. In *Proceedings of the 2016 ACM Workshop on Moving Target Defense*, pages 81–92. ACM, 2016.

179. Florian Mansmann, Timo Gbel, and William Cheswick. Visual Analysis of Complex Firewall Configurations. In *Proceedings of the 9th International Symposium on Visualization for Cyber Security*, pages 1–8. ACM, 2012.

180. Lockheed Martin. Cyber kill Chain®. *URL: http://cyber.lock-heedmartin.com/hubfs/Gaining_the_Advantage_Cyber_Kill_Chain.pdf*, 2014.

181. Sjouke Mauw and Martijn Oostdijk. Foundations of Attack Trees. In *Icisc*, volume 3935, pages 186–198. Springer, 2005.

182. Alain Mayer, Avishai Wool, and Elisha Ziskind. Fang: A Firewall Analysis Engine. In *Proceedings of the 2000 IEEE Symposium on Security and Privacy*, pages 177–187. IEEE, 2000.

183. McAfee. APT Protection Market Quadrant. https://www.mcafee.com/enterprise/en-us/assets/reports/rp-apt-market-quadrant-atd-top-player.pdf, 2017.

184. Todd McGuiness. Defense in Depth. *SANS Institute InfoSec Reading Room. SANS Institute*, 2001.

185. Nick McKeown, Tom Anderson, Hari Balakrishnan, Guru Parulkar, Larry Peterson, Jennifer Rexford, Scott Shenker, and Jonathan Turner. OpenFlow: Enabling Innovation in Campus Networks. *ACM SIGCOMM Computer Communication Review*, 38(2):69–74, 2008.

186. Marshall Kirk McKusick, George V Neville-Neil, and Robert NM Watson. *The Design and Implementation of the FreeBSD Operating System*. Pearson Education, 2014.

187. Ahmed M Medhat, Tarik Taleb, Asma Elmangoush, Giuseppe A Carella, Stefan Covaci, and Thomas Magedanz. Service Function Chaining in Next Generation Networks: State of the Art and Research Challenges. *IEEE Communications Magazine*, 55(2):216–223, 2017.

188. Jan Medved, Robert Varga, Anton Tkacik, and Ken Gray. Opendaylight: Towards a Model-Driven SDN Controller Architecture. In *2014 IEEE 15th International Symposium on*, pages 1–6. IEEE, 2014.

189. Mehta, Vaibhav, et al. "Ranking attack graphs." International Workshop on Recent Advances in Intrusion Detection. Springer, Berlin, Heidelberg, 2006.

190. Peter Mell, Karen Scarfone, and Sasha Romanosky. Common Vulnerability Scoring System. *IEEE Security & Privacy*, 4(6), 2006.

191. Dirk Merkel. Docker: Lightweight Linux Containers for Consistent Development and Deployment. *Linux Journal*, 2014(239):2, 2014.

192. Rashid Mijumbi, Joan Serrat, Juan-Luis Gorricho, Niels Bouten, Filip De Turck, and Raouf Boutaba. Network Function Virtualization: State-of-the-Art and Research Challenges. *IEEE Communications Surveys & Tutorials*, 18(1):236–262, 2016.

193. Jonathan D Moffett. Requirements and Policies. In *Proceedings of the Workshop on Policies for Distributed Systems and Networks (POLICY 1999)*, UK, 1999. HP Laboratories Bristol.

194. Jonathan D Moffett and Morris S Sloman. Policy Hierarchies for Distributed Systems Management. *IEEE Journal on Selected Areas in Communications*, 11(9):1404–1414, 1993.

195. Christopher Monsanto, Joshua Reich, Nate Foster, Jennifer Rexford, David Walker, and others. Composing Software-Defined Networks. In *Proceedings of the 10th USENIX Symposium on Networked Systems Design and Implementation (NSDI '13)*, pages 1–13. USENIX Association, 2013.

196. B. Moore. Policy Core Information Model (PCIM) Extensions. RFC 3460, IETF, January 2003.

197. Andrea Morgagni, Andrea Fiaschetti, Josef Noll, Ignacio Arenaza-Nuño, and Javier Del Ser. Security, Privacy, and Dependability Metrics. *Measurable and Composable Security, Privacy, and Dependability for Cyber-physical Systems: The SHIELD Methodology*, page 159, 2017.

198. Marti Motoyama, Damon McCoy, Kirill Levchenko, Stefan Savage, and Geoffrey M Voelker. An Analysis of Underground Forums. In *Proceedings of the 2011 ACM SIGCOMM Conference on Internet Measurement Conference*, pages 71–80. ACM, 2011.

199. John Moy. OSPF version 2. Technical report, 1997.

200. Marcelo Ribeiro Nascimento, Christian Esteve Rothenberg, Marcos Rogério Salvador, and Maurício Ferreira Magalhães. Quagflow: Partnering Quagga with Openflow. *ACM SIGCOMM Computer Communication Review*, 41 (4):441–442, 2011.

201. Janakarajan Natarajan. Analysis and Visualization of OpenFlow Rule Conflicts. Master's thesis, Arizona State University, 2016.

202. Sriram Natarajan, Xin Huang, and Tilman Wolf. Efficient Conflict Detection in Flow-Based Virtualized Networks. In *Proceedings of the 2012 International Conference on Computing, Networking and Communications (ICNC 2012)*, pages 690–696. IEEE, 2012.

203. Saran Neti, Anil Somayaji, and Michael E Locasto. Software Diversity: Security, Entropy and Game Theory.

204. NIST. NIST Software Defined Virtual Networks Project. available at http://searchsdn.techtarget.com/answer/What-is-the-difference-between-SDN-and-NFV, year.

205. Andres Ojamaa, Enn Tyugu, and Jyri Kivimaa. Pareto-Optimal Situaton Analysis for Selection of Security Measures. In *Military Communications Conference, 2008. MILCOM 2008. IEEE*, pages 1–7. IEEE, 2008.

206. OpenNFV. OpenNFV Based Service Function Chaining. https://wiki.opnfv.org/display/sfc/Service+Function+Chaining+Home, 2018. Online; accessed 13 February 2018.

207. OpenStack. available at https://www.openstack.org/.

208. International Standards Organisation. Intermediate System to Intermediate System Intra-Domain Routeing Exchange Protocol for use in Conjunction

with the Protocol for Providing the Connectionless-Mode Network Service (iso 8473). *ISO DP 10589*, February 1990.

209. Kouns, Jake. Open Source Vulnerability Database. The Open Source Business Resource (2008): 4.

210. Xinming Ou and Andrew W Appel. *A Logic-Programming Approach to Network Security Analysis*. Princeton University Princeton, 2005.

211. Xinming Ou, Sudhakar Govindavajhala, and Andrew W Appel. Mulval: A Logic-Based Network Security Analyzer. In *USENIX Security Symposium*, pages 8–8. Baltimore, MD, 2005.

212. Lawrence Page, Sergey Brin, Rajeev Motwani, and Terry Winograd. The Pagerank Citation Ranking: Bringing Order to the Web. Technical Report, Stanford InfoLab, 1999.

213. Justin Gregory V Pena and William Emmanuel Yu. Development of a Distributed Firewall Using Software Defined Networking Technology. In *Proceedings of the 4th International Conference on Information Science and Technology (ICIST 2014)*, pages 449–452. IEEE, 2014.

214. Ben Pfaff and Bruce Davie. The Open vSwitch Database Management Protocol. RFC 7047, IETF, 2013.

215. Ben Pfaff, Justin Pettit, Teemu Koponen, Ethan J Jackson, Andy Zhou, Jarno Rajahalme, Jesse Gross, Alex Wang, Joe Stringer, Pravin Shelar, et al. The Design and Implementation of Open Vswitch. In *NSDI*, pages 117–130, 2015.

216. Kvin Phemius, Mathieu Bouet, and Jrmie Leguay. DISCO: Distributed Multi-Domain SDN Controllers. In *Proceedings of the 2014 IEEE Network Operations and Management Symposium (NOMS 2014)*, pages 1–4. IEEE, May 2014.

217. David Lynton Poole, Alan K Mackworth, and Randy Goebel. *Computational intelligence: a logical approach*, volume 1. Oxford University Press New York, 1998.

218. Philip Porras, Seungwon Shin, Vinod Yegneswaran, Martin Fong, Mabry Tyson, and Guofei Gu. A Security Enforcement Kernel for Openflow Networks. In *Proceedings of the 1st Workshop on Hot Topics in Software Defined Networking (HotSDN 2012)*, pages 121–126. ACM, 2012.

219. Phillip A Porras, Steven Cheung, Martin W Fong, Keith Skinner, and Vinod Yegneswaran. Securing the Software Defined Network Control Layer. In *Proceedings of the Network and Distributed System Security Symposium 2015 (NDSS 15)*. ISOC, 2015.

220. Sergio Pozo, Rafael Ceballos, and Rafael M Gasca. AFPL, an Abstract Language Model for Firewall ACLs. In *Proceedings of the 8th International Conference on Computational Science and Its Applications (ICCSA 2008)*, pages 468–483. Springer, 2008.

221. Chaithan Prakash, Jeongkeun Lee, Yoshio Turner, Joon-Myung Kang, Aditya Akella, Sujata Banerjee, Charles Clark, Yadi Ma, Puneet Sharma, and Ying Zhang. Pga: Using graphs to express and automatically reconcile network policies. In *ACM SIGCOMM Computer Communication Review*, volume 45, pages 29–42. ACM, 2015.

222. Zafar Ayyub Qazi, Cheng-Chun Tu, Luis Chiang, Rui Miao, Vyas Sekar, and Minlan Yu. SIMPLE-fying Middlebox Policy Enforcement Using SDN. In *Proceedings of the 2013 ACM Conference on Special Interest Group on Data Communication (SIGCOMM '13)*, pages 27–38. ACM, 2013.

223. Paul Quinn, U Elzur, and C Pignataro. Network Service Header (nsh). Technical report, 2018.

224. Quinn, Paul, and Tom Nadeau. Problem Statement for Service Function Chaining. No. RFC 7498. 2015.

225. Jane Radatz, Anne Geraci, and Freny Katki. IEEE Standard Glossary of Software Engineering Terminology. *IEEE Std*, 610121990(121990):3, 1990.

226. David Raymond, Gregory Conti, Tom Cross, and Michael Nowatkowski. Key Terrain in Cyberspace: Seeking the High Ground. In *Proceedings of the 6th International Conference on Cyber Conflict (CyCon 2014)*, pages 287–300. IEEE, 2014.

227. Yakov Rekhter, Tony Li, and Susan Hares. A Border Gateway Protocol 4 (BGP-4). Technical report, 2005.

228. François Reynaud, François-Xavier Aguessy, Olivier Bettan, Mathieu Bouet, and Vania Conan. Attacks Against Network Functions Virtualization and Software-Defined Networking: State-of-the-Art. In *Proceedings of the 2016 IEEE Conference on Network Softwarization (NetSoft)*, pages 471–476. IEEE, 2016.

229. Andrew R Riddle and Soon M Chung. A Survey on the Security of Hypervisors in Cloud Computing. In *Distributed Computing Systems Workshops (ICDCSW), 2015 IEEE 35th International Conference on*, pages 100–104. IEEE, 2015.

230. D Romão, N Van Dijkhuizen, S Konstantaras, and G Thessalonikefs. Practical Security Analysis of Openflow. *University of Amsterdam, Amsterdam*, 2013.

231. Frank Rosenblatt. The Perceptron: A Probabilistic Model for Information Storage and Organization in the Brain. *Psychological Review*, 65(6):386, 1958.

232. Ronald S Ross. Managing Information Security Risk: Organization, Mission, and Information System View. *Special Publication (NIST SP)-800-39*, 2011.

233. Christian Esteve Rothenberg, Marcelo Ribeiro Nascimento, Marcos Rogerio Salvador, Carlos Nilton Araujo Corrêa, Sidney Cunha de Lucena, and Robert Raszuk. Revisiting Routing Control Platforms with the Eyes and Muscles of Software-Defined Networking. In *Proceedings of the First Workshop on Hot Topics in Software Defined Networks*, pages 13–18. ACM, 2012.

234. Arpan Roy. *Attack Countermeasure Trees: A Non-State-Space Approach Towards Analyzing Security and Finding Optimal Countermeasure Sets*. PhD thesis, Duke University, 2010.

235. Arpan Roy, Dong Seong Kim, and Kishor S Trivedi. Cyber Security Analysis Using Attack Countermeasure Trees. In *Proceedings of the Sixth Annual Workshop on Cyber Security and Information Intelligence Research*, page 28. ACM, 2010.

236. Arpan Roy, Dong Seong Kim, and Kishor S Trivedi. Attack Countermeasure trees (ACT): Towards Unifying the Constructs of Attack and Defense Trees. *Security and Communication Networks*, 5(8):929–943, 2012.

237. Arthur L Samuel. Some Studies in Machine Learning Using the Game of Checkers. *IBM Journal of Research and Development*, 3(3):210–229, 1959.

238. Wayne Sandholtz. Institutions and Collective Action: The New Telecommunications in Western Europe. *World Politics*, 45(2):242–270, 1993.

239. Fred B Schneider. Least Privilege and More. *IEEE Security & Privacy*, 1(5):55–59, 2003.

240. Bruce Schneier. Attack trees. *Dr. Dobbs Journal*, 24(12):21–29, 1999.

241. Sandra Scott-Hayward, Gemma O'Callaghan, and Sakir Sezer. SDN security: A survey. In *Future Networks and Services (SDN4FNS), 2013 IEEE SDN For*, pages 1–7. IEEE, 2013.

242. Omar Sefraoui, Mohammed Aissaoui, and Mohsine Eleuldj. Openstack: Toward an Open-Source Solution for Cloud Computing. *International Journal of Computer Applications*, 55(3), 2012.

243. Vyas Sekar, Sylvia Ratnasamy, Michael K Reiter, Norbert Egi, and Guangyu Shi. The Middlebox Manifesto: Enabling Innovation in Middlebox Deployment. In *Proceedings of the 10th ACM Workshop on Hot Topics in Networks (HotNets-X)*, page 21. ACM, 2011.

244. Alireza Shameli Sendi, Yosr Jarraya, Makan Pourzandi, and Mohamed Cheriet. Efficient Provisioning of Security Service Function Chaining Using Network Security Defense Patterns. *IEEE Transactions on Services Computing*, 2016.

245. Sezer, Sakir et al. Are We Ready for SDN? Implementation Challenges for Software-Defined Networks. *IEEE Communications Magazine*, 51.7, 2013, 36–43.

246. Hovav Shacham, Matthew Page, Ben Pfaff, Eu-Jin Goh, Nagendra Modadugu, and Dan Boneh. On the Effectiveness of Address-Space Randomization. In *Proceedings of the 11th ACM conference on Computer and communications security*, pages 298–307. ACM, 2004.

247. Susan J Shepard. Policy-Based Networks: Hype and Hope. *IT Professional*, 2(1):12–16, January 2000.

248. Justine Sherry, Shaddi Hasan, Colin Scott, Arvind Krishnamurthy, Sylvia Ratnasamy, and Vyas Sekar. Making Middleboxes Someone Else's Problem: Network Processing as a Cloud Service. *ACM SIGCOMM Computer Communication Review*, 42(4):13–24, 2012.

249. Sherry, Justine et al. Making Middleboxes Someone Else's Problem: Network Processing as a Cloud Service. *ACM SIGCOMM Computer Communication Review*, 42.4, 2012, 13–24.

250. Alaauddin Shieha. Application Layer Firewall Using OpenFlow. Master's thesis, University of Aleppo, 2014.

251. Seungwon Shin, Phillip A Porras, Vinod Yegneswaran, Martin W Fong, Guofei Gu, and Mabry Tyson. FRESCO: Modular Composable Security Services for Software-Defined Networks. In *Proceedings of the Network and Distributed System Security Symposium 2013 (NDSS 13)*. ISOC, February 2013.

252. Seungwon Shin, Vinod Yegneswaran, Phillip Porras, and Guofei Gu. Avant-Guard: Scalable and Vigilant Switch Flow Management in Software-Defined Networks. In *Proceedings of the 20th ACM Conference on Computer and Communications Security (CCS '13)*, pages 413–424. ACM, 2013.

253. Anirudh Sivaraman, Mihai Budiu, Alvin Cheung, Changhoon Kim, Steve Licking, George Varghese, Hari Balakrishnan, Mohammad Alizadeh, and Nick McKeown. Packet Transactions: A Programming Model for Data-plane Algorithms at Hardware Speed. *CoRR, vol. abs/1512.05023*, 2015.

254. Morris Sloman, Jeff Magee, Kevin Twidle, and J Kramer. An Architecture for Managing Distributed Systems. In *Proceedings of the 4th Workshop on Future Trends of Distributed Computing Systems*, pages 40–46. IEEE, September 1993.

255. Robin Sommer and Vern Paxson. Outside the Closed World: On using Machine Learning for Network Intrusion Detection. In *Proceedings of the 2010 IEEE Symposium on Security and Privacy*, pages 305–316. IEEE, 2010.

256. John Sonchack, Anurag Dubey, Adam J Aviv, Jonathan M Smith, and Eric Keller. Timing-based Reconnaissance and Defense in Software-Defined

Networks. In *Proceedings of the 32nd Annual Conference on Computer Security Applications*, pages 89–100. ACM, 2016.

257. Lance Spitzner. *Honeypots: Tracking Hackers*, volume 1. Addison-Wesley Reading, 2003.

258. William Stallings. SNMPv3: A Security Enhancement for SNMP. *IEEE Communications Surveys*, 1(1):2–17, 1998.

259. Stanford Open Flow Team. OpenFlow Switch Specification, Version 1.0.0. http://www.openflowswitch.org/documents/openflow-spec-v1.0.0.pdf, 2010.

260. John Strassner and Stephen Schleimer. Policy Framework Definition Language. Internet Draft, IETF, November 1998.

261. Michelle Suh, Sae Hyong Park, Byungjoon Lee, and Sunhee Yang. Building Firewall Over the Software-Defined Network Controller. In *Proceedings of the 16th International Conference on Advanced Communication Technology (ICACT2014)*, pages 744–748. IEEE, 2014.

262. Chen Sun, Jun Bi, Zhilong Zheng, Heng Yu, and Hongxin Hu. NFP: Enabling Network Function Parallelism in NFV. In *Proceedings of the Conference of the ACM Special Interest Group on Data Communication*, pages 43–56. ACM, 2017.

263. Tacker. Openstack Based SFC. https://wiki.openstack.org/wiki/Tacker, 2018. Online; accessed 13 February 2018.

264. Arsalan Tavakoli, Martin Casado, Teemu Koponen, and Scott Shenker. Applying NOX to the Datacenter. In *Proceedings of the 8th ACM Workshop on Hot Topics in Networks (HotNets-VIII)*. ACM, 2009.

265. Mahbod Tavallaee, Ebrahim Bagheri, Wei Lu, and Ali A Ghorbani. A Detailed Analysis of the KDD CUP 99 Data Set. In *Computational Intelligence for Security and Defense Applications, 2009. CISDA 2009. IEEE Symposium on*, pages 1–6. IEEE, 2009.

266. Mark Thompson, Noah Evans, and Victoria Kisekka. Multiple OS Rotational Environment an Implemented Moving Target Defense. In *Proceedings of the 7th International Symposium on Resilient Control Systems (ISRCS 2014)*, pages 1–6. IEEE, 2014.

267. Amin Tootoonchian and Yashar Ganjali. HyperFlow: A Distributed Control Plane for OpenFlow. In *Proceedings of the 2010 Internet Network Management Workshop/Workshop on Research on Enterprise Networking (INM/WREN '10))*, pages 3–3. USENIX Association, 2010.

268. Irena Trajkovska, Michail-Alexandros Kourtis, Christos Sakkas, Denis Baudinot, João Silva, Piyush Harsh, George Xylouris, Thomas Michael Bohnert, and Harilaos Koumaras. SDN-based Service Function Chaining Mechanism and Service Prototype Implementation in NFV Scenario. *Computer Standards & Interfaces*, 54:247–265, 2017.

269. Tung Tran, Ehab S Al-Shaer, and Raouf Boutaba. PolicyVis: Firewall Security Policy Visualization and Inspection. In *LISA*, volume 7, pages 1–16, 2007.

270. Enn Tyugu. Artificial Intelligence in Cyber Defense. In *Cyber Conflict (ICCC), 2011 3rd International Conference on*, pages 1–11. IEEE, 2011.

271. Nuutti Varis. Anatomy of a Linux Bridge. In *Proceedings of Seminar on Network Protocols in Operating Systems*, page 58, 2012.

272. Vijay V Vazirani. *Approximation Algorithms*. Springer Science & Business Media, 2013.

273. Patrick Verkaik, Dan Pei, Tom Scholl, Aman Shaikh, Alex C Snoeren, and Jacobus E Van Der Merwe. Wresting Control from BGP: Scalable Fine-Grained Route Control. In *USENIX Annual Technical Conference*, pages 295–308, 2007.

274. Dinesh C Verma. Simplifying Network Administration Using Policy-Based Management. *IEEE Network*, 16(2):20–26, 2002.

275. VMWare. VMWare NSX: Network Virtualization and Security Platform. https://www.vmware.com/products/nsx.html, 2018.

276. Liberios Vokorokos, Anton Balaz, and Martin Chovanec. Intrusion Detection System using Self Organizing Map. *Acta Electrotechnica et Informatica*, 6(1):1–6, 2006.

277. Huazhe Wang, Xin Li, Yu Zhao, Ye Yu, Hongkun Yang, and Chen Qian. SICS: Secure In-Cloud Service Function Chaining. *arXiv preprint arXiv:1606.07079*, 2016.

278. Gary Williams. Operations Security (OPSEC). *Ft. Leavenworth, Kan.: Center for Army Lessons Learned*, 1999.

279. Steve Williams. The Softswitch Advantage. *IEE Review*, 48(4):25–29, 2002.

280. Wald Wojdak. Rapid Spanning Tree Protocol: A New Solution from an Old Technology. *Reprinted from CompactPCI Systems*, 2003.

281. G. H. von Wright. Deontic Logic. *Mind*, 60(237):1–15, 1951.

282. Peng Xiao, Wenyu Qu, Heng Qi, Zhiyang Li, and Yujie Xu. The SDN Controller Placement Problem for WAN. In *Proceedings of the 2014 IEEE/CIC International Conference on Communications in China (ICCC)*, pages 220–224. IEEE, 2014.

283. Jun Xu, Pinyao Guo, Mingyi Zhao, Robert F Erbacher, Minghui Zhu, and Peng Liu. Comparing Different Moving Target Defense Techniques. In *Proceedings of the 1st ACM Workshop on Moving Target Defense (MTD 2014)*, pages 97–107. ACM, 2014.

284. Justin Yackoski, Harry Bullen, Xiang Yu, and Jason Li. Applying Self-Shielding Dynamics to the Network Architecture. In *Moving Target Defense II*, pages 97–115. Springer, 2013.

285. Zheng Yan and Christian Prehofer. Autonomic Trust Management for a Component-based Software System. *IEEE Transactions on Dependable and Secure Computing*, 8(6):810–823, 2011.

286. Lily Yang, Ram Dantu, Terry Anderson, and Ram Gopal. Forwarding and Control Element Separation (ForCES) Framework. Technical report, 2004.

287. Wei Yang and Carol Fung. A Survey on Security in Network Functions Virtualization. In *Proceedings of the 2016 IEEE Conference on Network Softwarization (NetSoft)*, pages 15–19. IEEE, 2016.

288. Guang Yao, Jun Bi, Yuliang Li, and Luyi Guo. On the Capacitated Controller Placement Problem in Software-Defined Networks. *IEEE Communications Letters*, 18(8):1339–1342, August 2014.

289. Soheil Hassas Yeganeh and Yashar Ganjali. Kandoo: A Framework for Efficient and Scalable Offloading of Control Applications. In *The Beacon Openflow Controller*, pages 19–24. ACM, 2012.

290. Soheil Hassas Yeganeh, Amin Tootoonchian, and Yashar Ganjali. On Scalability of Software-Defined Networking. *IEEE Communications Magazine*, 51(2):136–141, 2013.

291. Minlan Yu, Jennifer Rexford, Michael J Freedman, and Jia Wang. Scalable Flow-Based Networking with DIFANE. *ACM SIGCOMM Computer Communication Review*, 40(4):351–362, 2010.

292. Wei Yu, Xinwen Fu, Steve Graham, Dong Xuan, and Wei Zhao. Dsss-based Flow Marking Technique for Invisible Traceback. In *Proceedings of the 2007 IEEE Symposium on Security and Privacy*, pages 18–32. IEEE, 2007.

293. Lihua Yuan, Hao Chen, Jianning Mai, Chen-Nee Chuah, Zhendong Su, and Prasant Mohapatra. Fireman: A Toolkit for Firewall Modeling and Analysis. In *Proceedings of the 2006 IEEE Symposium on Security and Privacy*, pages 15–29. IEEE, 2006.

294. Kara Zaffarano, Joshua Taylor, and Samuel Hamilton. A Quantitative Framework for Moving Target Defense Effectiveness Evaluation. In *Proceedings of the Second ACM Workshop on Moving Target Defense*, pages 3–10. ACM, 2015.

295. Bin Zhang, Ehab Al-Shaer, Radha Jagadeesan, James Riely, and Corin Pitcher. Specifications of a High-Level Conflict-Free Firewall Policy Language for Multi-Domain Networks. In *Proceedings of the 12th ACM Symposium on Access Control Models and Technologies (SACMAT '07)*, pages 185–194. ACM, 2007.

296. Quanyan Zhu and Tamer Başar. Game-theoretic Approach to Feedback-driven Multi-stage Moving Target Defense. In *International Conference on Decision and Game Theory for Security*, pages 246–263. Springer, 2013.

297. Rui Zhuang, Scott A DeLoach, and Xinming Ou. Towards a Theory of Moving Target Defense. In *Proceedings of the 1st ACM Workshop on Moving Target Defense (MTD 2014)*, pages 31–40. ACM, 2014.

298. Rui Zhuang, Su Zhang, Alex Bardas, Scott A DeLoach, Xinming Ou, and Achintya Singhal. Investigating the Application of Moving Target Defenses to Network Security. In *Proceedings of the 6th International Symposium on Resilient Control Systems (ISRCS 2013)*, pages 162–169. IEEE, 2013.

Index

Printed in the United States
by Baker & Taylor Publisher Services